T0342118

The Evolution of Scientific Knowledge

NEW HORIZONS IN INSTITUTIONAL AND EVOLUTIONARY ECONOMICS

**Series Editor:** Geoffrey M. Hodgson
Research Professor, University of Hertfordshire Business School, UK

Economics today is at a crossroads. New ideas and approaches are challenging the largely static and equilibrium-oriented models that used to dominate mainstream economics. The study of economic institutions – long neglected in the economics textbooks – has returned to the forefront of theoretical and empirical investigation.

This challenging and interdisciplinary series publishes leading works at the forefront of institutional and evolutionary theory and focuses on cutting-edge analyses of modern socio-economic systems. The aim is to understand both the institutional structures of modern economies and the processes of economic evolution and development. Contributions will be from all forms of evolutionary and institutional economics, as well as from Post-Keynesian, Austrian and other schools. The overriding aim is to understand the processes of institutional transformation and economic change.

Titles in the series include:

The New Evolutionary Microeconomics
Complexity, Competence and Adaptive Behaviour
*Jason D. Potts*

National Competitiveness and Economic Growth
The Changing Determinants of of Economic Performance in the World Economy
*Timo J. Hämäläinen*

Conventions and Structures in Economic Organization
Markets, Networks and Hierarchies
*Edited by Olivier Favereau and Emmanuel Lazega*

Globalization and Institutions
Redefining the Rules of the Economic Game
*Edited by Marie-Laure Djelic and Sigrid Quack*

The Evolutionary Analysis of Economic Policy
*Edited by Pavel Pelikan and Gerhard Wegner*

The Evolution of Scientific Knowledge
*Edited by Hans Siggaard Jensen, Lykke Margot Richter and Morten Thanning Vendelø*

# The Evolution of Scientific Knowledge

*Edited by*

Hans Siggaard Jensen and Lykke Margot Richter

*Learning Lab Denmark*

Morten Thanning Vendelø

*Copenhagen Business School, Denmark*

NEW HORIZONS IN INSTITUTIONAL AND EVOLUTIONARY ECONOMICS

**Edward Elgar**

Cheltenham, UK • Northampton, MA, USA

Published by
Edward Elgar Publishing Limited
Glensanda House
Montpellier Parade
Cheltenham
Glos GL50 1UA
UK

Edward Elgar Publishing, Inc.
136 West Street
Suite 202
Northampton
Massachusetts 01060
USA

A catalogue record for this book
is available from the British Library

**Library of Congress Cataloguing in Publication Data**
The evolution of scientific knowledge / edited by Hans Siggaard Jensen, Lykke Margot Richter, Morten Thanning Vendelø.
p. cm. — (New Horizons in institutional and evolutionary economics series)
Includes bibliographical references.
1. Science—Philosophy. 2. Science—History. 3. Science—Social aspects. I. Jensen, Hans Siggaard, 1947– II. Richter, Lykke Margot, 1965– III. Vendelø, Morten Thanning, 1976– IV. New Horizons in institutional and evolutionary economics.

Q175.E93 2003
501—dc21

2003040773

ISBN 1 84376 235 8

Printed and bound in Great Britain by MPG Books Ltd, Bodmin, Cornwall

# Contents

# Figures



# Tables

# Contributors

**William N. Butos**, Professor of Economics at Trinity College, Hartford, Connecticut

**Finn Collin**, Professor of Philosophy, Department of Education, Philosophy and Rhetoric, University of Copenhagen, Denmark

**Mathew Forstater**, Associate Professor, Department of Economics, University of Missouri-Kansas City, Missouri

**Hans Siggaard Jensen**, Professor and Research Director, Learning Lab Denmark

**Jukka Kaisla**, Research Associate, Department of Economics, Helsinki School of Economics, Finland

**Matthias Klaes**, Lecturer, Department of Economics, University of Stirling, United Kingdom

**Christian Knudsen**, Professor, Department of Industrial Economics and Strategy, Copenhagen Business School, Denmark

**Thorbjørn Knudsen**, Associate Professor, Department of Marketing, University of Southern Denmark, Odense Campus

**Roger Koppl**, Professor of Economics and Finance at Fairleigh Dickinson University, Madison, New Jersey

**Laurence S. Moss**, Professor of Economics at Babson College, Boston, Massachusetts

**Lykke Margot Richter**, Research Associate, Learning Lab Denmark

**Morten Thanning Vendelø**, Associate Professor, Head of Department, Department of Informatics, Copenhagen Business School, Denmark

# Preface

The event of this book takes its departure in the workshop 'Science as a Spontaneous Order' 2000. It all began on a cold winter day when we and the participants from England, Finland, Germany, Holland, Sweden and the USA set off in two minibuses with the destination: Klitgaarden, Skagen in Denmark. The occasion was the inauguration of the newly restored King Christian X's summer 'cottage', which had been turned into a refugium for scientists and artists. The city of Skagen is situated at the northernmost tip of Denmark, and it is famous for the unique light created by reflection of the sunlight in the two seas surrounding the city, as portrayed by some of the famous Skagen painters.[1]

The theme for the workshop was grounded in a desire to use the first workshop at Klitgaarden Refugium to focus on science as a phenomenon and on the patterns of development in science. Traditionally, philosophers and historians have analysed science as an institution; however within the last decades, especially, economists have shown an increasing interest in institutions, and furthermore, an increasing interest in the use of the evolutionary perspective in understanding the emergence and evaporation of institutions. Science of science is thus an example of a field where sociology, philosophy and economics meet in an attempt to elaborate on the complexity of the emergence and functions of organisations.

The workshop included some of the leading scientists within these disciplines: sociology of science, philosophy of science, economics of science and history of science. Gathering scientists from these various disciplines meant that patrons of F. A. von Hayek's idea of the economic information system as 'a spontaneous order' faced the challenge of arguing against the constructivist position. During the workshop there was time to enjoy the, in January, barbarian landscape of the northernmost tip of Denmark, which in contrast made the tense debate look like a peaceful fight at the Coliseum.

The workshop was sponsored by Aalborg University and Copenhagen Business School, and we thank our sponsors for their generosity, which led to the initiative of inviting a number of authors to contribute to this volume. The papers written by authors agreeing to contribute to the volume were subsequently reviewed by us, the editors, and revised accordingly. Above all, this happened in an attempt to give the papers a stronger focus on scientific knowledge, and therefore it made sense for us as the editors to position the following contributions of this book as dealing with the evolution of scientific knowledge, and so it acquired in our and our publisher's opinion a more snappy title.

Our thanks to the contributors to this volume, as without their warm-hearted and patient collaboration, we would not have been able to turn an impassioned idea into a book.

## NOTES

1 Danish painters of Skagen: Sören Kröyer (1851–1909), Lauritz Tuxen (1853–1927), Anna Ancher (1859–1935) and Michael Ancher (1849–1927).

# 1. Introduction

**Hans Siggaard Jensen, Lykke Margot Richter and Morten Thanning Vendelø**

---

Over the past decades science as an institution has been analysed from different perspectives. Scientific revolutions attached to names like Copernicus, Newton, Einstein and Bohr have resulted in new discoveries about the world and also led to new inventions and vice versa: new inventions or discoveries led to scientific revolution. A scientific revolution is characterised by the scientific community rejecting old theories and replacing them with new, to such an extent that the new way of looking at the world is incompatible with the old way of looking at the world (Kuhn 1970).

While the history of science provides a picture of how empirical knowledge contributes to science, the philosophy of science plays an important role in understanding what scientific knowledge is. To fully understand essential features of science we need to study how personal ideas, models and concepts have emerged, and we need to study how they have developed into scientific knowledge and what within a scientific community makes a good theory.

In our opinion the question cannot be answered without a notion of philosophical scepticism. This notion has previously been presented by, for example, Cohen (1977, pp. 308–49) and later Redman (1991). Besides the historical and the more traditional philosophical perspectives on science, science has both been viewed as a result of social constructions and it has been perceived as an important economic factor in the so-called 'knowledge society'. Hence, philosophical understanding of science has recently been influenced by new perspectives of sociological and historical work on science.

The purpose of this book is to provide the reader with an understanding of science and of the research system both at the ontological level and at the empirical level. The economic theories used are of the institutional and evolutionary variety, and the sociological theories draw from the type of work on social studies of science that have in the last decades transformed our picture of science and technology. In order to form a body of ideas and techniques of scientific knowledge, we look at science from an evolutionary perspective. Science is about principles, evidence and methods.

The study of why some theories win over others implies the rationality and motivation of the science system and includes economics of science. The question therefore calls, maybe not for explanations, but for the ability to show some of the complexity in science.

Within the last decades, economics has become increasingly interested in institutions and organisations, and in applying several perspectives, for example, evolutionary and Austrian, to the emergence and functioning of institutions (besides the more traditional ways of understanding, exemplified by the neo-classical orthodoxy). Furthermore, the role of power has been emphasised. In the history of science we find many examples of scientific results not published until long after they were explored or even after the death of the scientist, simply because alternatives were too dangerous for those in power, as in the case of the Copernican world view (Kuhn 1957).

Science is not a simple concept nor is it a system. And every time we choose one perspective to look at science, we reveal something, while we hide something else.

So in this aspect, we draw the conclusion that many perspectives and approaches are necessary in order to understand science as a phenomenon. To emphasise the complexity we have developed six categories of understanding science – categories that are without any doubt neither definitive or inclusive, but accentuate the ambiguity:

- Science can be understood as the use of certain methods and practices or as a way of rule-following – an evolutionary selection.
- Science can be understood as a search for the truth. A concept that is deeply rooted in realism, here there reigns a positivistic world view claiming the world is really as our best scientific theories describe it.
- Science can be understood in an instrumental perspective – often attached to Ernst Mach's motto: 'Science is the economy of thoughts'.
- Science can be understood as power – 'knowledge is power' – and the conflict between scientists and politicians, and between scientists can be seen as a struggle for desired outcomes.
- Science can be understood as paradigms – where theories about the real world are seen as social constructions – both in our way of reconstructing the past and of constructing the future.

- Science can be understood as a process of learning – where the trial and error method serves as a basis for building theories on experience and falsification.

In the following sections, we first present science on science as a discipline by presenting the reflection of different disciplines on science as well as on the evolution of science. We then move on to discuss research programmes as well as individualism in these. Thereafter we look at the rationality of science and how competing rationalities may affect the process of scientific discovery; and finally, we discuss orders of science and their emergence.

## 1. SCIENCE ON SCIENCE

By tradition economists are interested in science, and 'the foremost reason economists have for studying science is the link between science and economic growth. That such a relationship exists has long been part of the conventional wisdom, articulated first by Adam Smith ([1776] 1982, p. 113)' (Stephan 1996: 1226). Over time this interest has evolved into a number of distinct fields of interest, each with their own research agenda. Stephan (1996: 1199) describes three major reasons for economists to be interested in science. First, the economic impact of science is indisputable, as it is evident that science provides for economic growth and welfare. Second, scientific labour markets are a fertile ground for study; and third, the reward structure that has evolved in science goes a long way in solving the appropriability problem associated with the production of a public good. Scientific results are produced in at least two different markets; university research and the research produced in industry. Furthermore, we have the distinction between basic research and applied research. From an economics of science perspective, it is a lot easier to estimate the value of science on the market, because the science attached to industry in, for example, technology, biology or other products is immediately beneficial, whereas basic research may have a life-cycle of forty years before we can point out the economic consequences.

In scientific reflection on science we find an example of a field where economics, sociology, and philosophy meet in attempts to understand the evolution of scientific knowledge.

## 2. EVOLUTION OF KNOWLEDGE

The evolutionary perspective on science originates from biology. The general notion of evolution is that from a variation of ideas, species, characteristics, and features, some are selected by environmental pressure, which may be the competition from other species or the preferences of potential mates, and retained as long as the environmental pressure selecting them is in place. Hence, the theme in evolutionary theory is that of variation, selection, and retention (Campbell [1974] 1987).

Scientists operating in a variety of fields such as economics and sociology have adopted this perspective. In the evolutionary perspective a realist approach has a central position, and thus, this perspective acknowledges the existence of an external world apart from our thoughts, about which we can obtain some knowledge. Hence, this perspective emphasises the importance of the pursuit of truth (Hodgson 1993, p. 10). Thus, in the evolutionary perspective 'a science is normally defined as the study of a particular aspect of objective reality: physics is about the nature and properties of matter and energy, biology about living things, psychology about the psyche, and so on' (Hodgson 1993, p. 7).

One of these perspectives on the evolution of science is Popper's theory of falsification. Popper was one of the first philosophers of science to break the positivistic picture of science. He took up Hume's critique of induction showing that verification of a scientific theory is utopia, because prediction of the future based on constellations of the past possibly leads to a false judgment. It can never be based on valid logical inference. Hence, Popper sets the demarcation criterion of scientific knowledge as its being falsifiable. This naïve falsificationism leads to fallibilism in the strong case, and in order to avoid such an entrapment, Popper stressed a regulative idea of an objective world connected to the correspondence criterion of truth. With the demarcation criterion he set the standards for what one could call 'the survival' of scientific theories and on the basis of this, he developed an evolutionary account of scientific knowledge.

Also, Kuhn's (1970) theory of the structure of scientific revolutions has evolutionary features. Kuhn (1970) opens up at least three phases in scientific progress: The phase of normal science; crisis; and scientific revolution. Hence, he focused more on the research community, the so-called scientific paradigm, consisting of social elements, like traditions, peers, and consensus.

In switching from one paradigm to another, there are no guarantees of knowledge-progress, because of the incommensurability of paradigms. (However, seen in a pragmatic-instrumental perspective, a later theory can prove to be more useful in solving 'puzzles'.) And this brings us to the critique of using the evolutionary model.

An evolutionary perspective will show us, in choosing between research programmes, theories and hypotheses, which ones are the 'fittest', but the fittest theories may not be the ones that are the best, neither truer nor relatively better. 'Only the successful theories survive – the ones which in fact have latched on to actual regularities in nature' (Van Fraassen 1980, p. 40 in Psillos 1999, p. 96); is that because of beauty or justified beliefs?

## 3. RESEARCH PROGRAMMES

The choice of research programmes and when to change research programmes are issues of central importance to scientists. In Chapter 2, C. Knudsen discusses this issue by focusing on the organisation of research. In the choice between research programmes, the competitiveness of science is split between strengthening a research programme and new innovative research programmes. The problem with research programmes is when they are used without scepticism or with too much scepticism. Since the 1960s, especially within the social sciences, we have experienced a boom in new theories, piles of books on management theories, organisational behaviour and so on, which are all indications of the race among scientists in coming up with new innovative research programmes and strategies. Coping with this 'theoretical jungle' implies time to absorb old theories, so that old theories are replaced and not just forgotten, and in order to avoid new theories simply creating fragmentation. In other words, if it is possible in the light of the incommensurability of scientific paradigms, we need old theories to serve as a solid knowledge base for the creation of innovative research programmes.

Kuhn (1970) explained stability in research programmes with reference to the social elements of a scientific paradigm, where traditions, habits and community practice form a joint world view. New theories are seen as the outcome of a revolution, a so-called change in world views. In contrast, Lakatos emphasised the heuristics of theories as guiding the research programmes, where the 'hard core' is seen as the more stable element and is supplemented by the 'protective belt'. In Chapter 3, Klaes addresses the problem of how community practice influences research programmes. However, he himself makes a paradigm shift by arguing that the scientist's manipulation of the 'hard core' can be seen as a spontaneous discourse in explanatory categories, and therefore, be productive in research programmes rather than have these resting on their laurels.

And so we are led into constructivism. According to Latour (1987) and Collins (1981) (and other believers in the strong programme) theories do not reflect reality, rather they reflect our perception of reality, and so we cannot know whether objective reality has any impact on our scientific beliefs. This perspective that can be split into at least two positions, epistemological and ontological, where the mantra of the latter is that scientific theories generate what is real, and where the claim of the epistemological position is that a sharp distinction between the context of discovery and the context of justification is impossible, so that there is no special role for objective reality in the context of justification. Collin in Chapter 4 discusses these two models in order to emphasise the weaknesses and strengths of the two different perspectives. However, the analysis of the two models reveals no winner; moreover the analysis seems to lead to an important philosophical point, namely that in order for a model or theory to take 'itself' seriously it must be self-adaptive, which means that in a reflective manner the theory also applies to itself.

## 4. INDIVIDUALISM IN RESEARCH PROGRAMMES

Studies of science suggest that science takes time. 'Investigators often portray productive scientists – and eminent scientists especially – as strongly motivated, with the "stamina" or the capacity to work hard and persist in the pursuit of long-range goals (Fox 1983: 287)' (Stephan 1996: 1219).

Focusing on how individuals arrive at new scientific insights, T. Knudsen, in Chapter 5, shows how implicit conceptions acquired at an early stage of a research programme can block new discoveries until modification of the implicit conceptions has happened. Using the case of Darwin's delay in publishing *The Origin of the Species by Means of Natural Selection*, he suggests that the process of scientific knowledge-creation is a complicated individualistic process, which demands a fair amount of time and social interaction with peers in the trade.

In the case of Darwin, T. Knudsen demonstrates that he 'needed time to develop the necessary implicit knowledge that could support his explicit selection principle'. T. Knudsen thus resists the idea that scientific discovery involves the conversion of implicit to explicit knowledge. Rather it is the acquisition of implicit knowledge that enables scientists to reflect on and elaborate their understanding of previously developed theoretical concepts. Thereby, it is suggested that, 'explicit codified knowledge must always be supported by uncodified implicit knowledge'. Using this observation as his point of departure T. Knudsen establishes a neo-Darwinian science model whose inheritance track is constituted by implicit and explicit knowledge operating in two separate but connected trails. He suggests that 'Darwin's absorption of explicit codified knowledge through literary contact benefited from the implicit knowledge he had absorbed through personal contact'. Therefore, scientific progress cannot happen without recurrent and intimate interaction, as scientific progress is achieved in interplay between individual absorption of explicit knowledge and social interaction, where shared cognitive categories and thereby tacit knowledge are developed.

In this process it is the development of shared cognitive categories that facilitates the exchange of implicit or tacit knowledge among the interacting parties (Denzau and North 1994). This points to the specific scientific community in which the scientist is embedded as constituting the selective and retentive environment for her or his scientific ideas. Ideas are likely to be conceptualised by individuals, but their further development depends on the social interaction with peers. In such environments shared cognitive categories make the participants more receptive to each other's ideas, ways of thinking and argumentations. A more intimate atmosphere evolves and facilitates intense exchange and development of tacit knowledge.

## 5. RATIONALITY OF SCIENCE

The questions concerning the rationalities guiding scientific work are of interest to historians of science. Addressing some of these issues Moss in Chapter 6 investigates the characteristics of scientific contests and their winner-take-all nature. In winner-take-all contests much is at stake: fame, recognition, awards, social surplus, and promotions, whereas few care about the runner-up, as the priority of discovery is a form of property right. The issue addressed by both Moss and later Kaisla in Chapter 7 can be read as if there exists such a thing as rules of good conduct in science. But what is the role of these rules of good conduct?

From the two cases presented by Moss it appears that rules of good conduct in science seem to be sufficiently ambiguous to allow for interpretation, and that different sets of rules exist simultaneously. The MacLeod-Banting case describes a situation where it is the research organisation, and embeddedness of the rules for assignment of ownership to scientific discoveries, that somehow enabled the scientific community to benefit from collective action and compromised the values and interests of Banting, who clearly perceived himself as the sole rightful owner of the insulin discovery.

The interesting question thus raised by constitutional theory is: was there a way in which the team-based research could have been organised to prevent the controversy that emerged, or do such contradictory positions as those held by Banting and MacLeod inherently result in conflicts over intellectual ownership? Kaisla discusses the contribution of constitutional theory to considerations regarding the organisation of scientific activities.

In our interpretation, the question addressed by Kaisla is whether or not science can be organised, such that the scientific community can harvest collective benefits without compromising the variety of values and interests held by the participating individuals. The core point of view is that voluntary exchange is the prime motivator for individuals to allow constraints to limit their behaviour. The MacLeod-Banting case demonstrates the value of the discussion of rules and rule-following. It is obvious that the two isolators of insulin did not subscribe to the same rules regarding authentication of scientific discoveries. They came from two different environments and held incompatible understandings of how propriety of scientific discoveries should be assigned. This interpretation is confirmed by Bliss (1982) who refers to the 'cowboy' or lone-rider mentality of Banting, who neglected the contribution of MacLeod in orchestrating the team-based research effort. On the contrary MacLeod tacitly adhered to emerging values emphasising the importance of collective scientific efforts. Under certain circumstances such controversies between a newcomer and the establishment may change the prevailing order, yet when the institutions of the establishment are strong, change is likely to be resisted.

The problem regarding rationalities of science appears to be the tacit aspect of their nature, because when rules for good conduct in science are predominantly implicit, and thus taken for granted by the actors, the outcome becomes unforeseeable, and therefore it resembles a spontaneous emergence. Probably the only possible predictor of the outcome is the strength of the institutions backing the various rationalities. In the MacLeod-Banting case that institution can be viewed to be the emerging perception of scientific discovery as resulting from a collectively orchestrated effort.

The argument put forward in the chapter by Moss appears to be that the MacLeod-Banting case could not have been designed, and could not even have been foreseen. The two sets of rationalities coming together were implicit, and did not surface until after actions had been taken and the scientific discoveries had been made. The same appears to be true in the case describing the Nobel Prize-winning scientific work carried out by Watson, Crick and Wilkins, for in the discovery of the DNA double helix their behaviour was more in concert with one another in their violation of the existing unwritten rules of British professionalism.

## 6. ORDERS OF SCIENCE

Orders of science are connected to the rules of rationalities within the different scientific disciplines. The tacit rationality of science may be phrased as a question of orders and how these emerge, and so Butos and Koppl in

Chapter 8 look at science as an emerging system, dominated by the distinction between designed and spontaneous order. The concept of spontaneous order originates from the Austrian Nobel Prize winner Friedrich A. von Hayek, who borrowed it both from Adam Smith's notion of 'the invisible hand' and from the Scottish natural law philosophers who argued that society developed into a spontaneous order, which was the result of human action but not human design. Hayek (1973) distinguishes between two types of order, those made by design and those grown spontaneously, and he defines order as (p. 36):

> ... a state of affairs in which a multiplicity of elements of various kinds are so related to each other that we may learn from our acquaintance with some spatial or temporal part of the whole to form correct expectations concerning the rest, or at least expectations which have a good chance of proving correct.

Some order, consistency or constancy always exists in social life. If this was not the case we would all face difficulties in going about our affairs or satisfying our most elementary needs. About this Hayek (1973, p. 36) says:

> Living as members of society and dependent for the satisfaction of most of our needs on various forms of co-operation with others, we depend for the effective pursuit of our aims clearly on the correspondence of the expectations concerning the actions of others on which our plans are based with that they will really do.

Hence, as orders produce rules and institutions, such as the market, they provide for coordination among the various specialised members of society. Designed orders 'rest on a relation of command and obedience, or a hierarchical structure of the whole of society in which the will of superiors, and ultimately of some single supreme authority, determines what each individual must do' (Hayek 1973, p. 36). However, managing knowledge by design implies two difficulties: (a) knowledge is often difficult to articulate Hayek (1948) and structured in an imperfect way (Langlois 1984); and (b) the notion of design tends to ignore the constantly changing circumstances under which knowledge is in use, and thus when managing knowledge by design the ever-changing environment is neglected. When viewing orders as emerging in a spontaneous way Butos and Koppl subscribe to 'the discovery that there exist orderly structures which are the product of the action of many men, but are not the result of human design' (Hayek 1973, p. 37). Hence, these structures are the result of evolutionary processes that nobody foresaw or designed. About the very nature of spontaneous orders Hayek (1973, p. 38) explains:

> Spontaneous orders are not necessarily complex, but unlike deliberate human arrangements, they may achieve any degree of complexity. One of our main contentions will be that very complex orders, comprising more particular facts than any brain could ascertain or manipulate, can be brought about only through forces inducing the formation of spontaneous orders.

Forstater in Chapter 9 elaborates on this; what once was spontaneous and unintended can later become intended and vice versa. It is a must if we choose to see science as not only a progress in the growth of knowledge, but as having an educational effect on the scientist in the way that science is about learning about our concepts of reality.

Jensen and Richter in Chapter 10 discuss different types of orders in science in relation to types of knowledge, as not only do different types of order result from different disciplines, but different types of knowledge can also be linked to the scientific process. We have embodied knowledge, disembodied knowledge, tacit knowledge, and implicit/explicit knowledge, which emphasises the fact that something substantial is at stake, whereas knowing seems to be attached to a process of personal experience.

In 1966 Michael Polanyi claimed that the rationality of science consists of tacit knowledge (Polanyi 1966), a claim that later was supported by experimental psychology (see T. Knudsen in Chapter 5). If so, it is clear that the concept will create problems, because 'tacit knowledge' is already secured as knowledge, but it is not structured like propositions or facts. Jensen and Richter claim that its disposition is more like a speech act, and making it explicit can be seen as creating a new set of typologies and rules of action. In this chapter the concepts of Modus 1 and Modus 2 in the production of knowledge (Gibbons et al. 1994) is related to 'tacit knowing' as the more precise interpretation of Polanyi, which leads to the conclusion that the type of work connected with producing knowledge in Modus 2 has features relating it to the embedded and situated position that agents find themselves in when acting in a market; while one could say that the knowledge of Modus 1 is like the knowledge economists get about market-rationality through investigation of models of the market.

## 7. CONCLUSION

In conclusion these different perspectives seem to point to science as neither an institution nor an order that emerged as the result of conscious and wilful design, nor as just like a 'normal' market. Science has aspects of market orders and aspects of orders created by design. Furthermore, science is a development that in some ways is like the development of economic systems, in other ways very different. With the focus on evolution, science faces new challenges. Many of these are interdisciplinary in nature, and involve ideas from economics, organisation theory, biology, cognitive psychology, and even physics.

So, the authors in this book explore aspects of economic evolutionary thinking in science at a foundational level. This includes topics such as the differences between the behavioural assumptions and the units of analysis used in the different theories; understanding innovations, and the evolution of knowledge; and different approaches to 'rationality' within the different theories. Questions dealt with include the methodological and philosophical background of evolutionary theories, dynamic theories of social institutions, and evolutionary understanding of markets and firms. In this way, the book explicitly addresses questions relating to the foundations of evolutionary thinking, on both the methodological, conceptual and theoretical levels. So from the bottom of our hearts, dear reader, we wish you a pleasant journey into our collection of science on science.

## REFERENCES

Bliss, Michael (1982), *The Discovery of Insulin*, Chicago, IL: University of Chicago Press.

Campbell, Donald T. ([1974] 1987), 'Evolutionary Epistemology', in Gerard Radnitsky and William W. Bartley III (eds), *Evolutionary Epistemology, Theory of Rationality, and the Sociology of Knowledge*, La Salle, IL: Open Court, pp. 47–89.

Cohen, Bernard I. (1977), 'History and the Philosophy of Science', in Frederick Suppe (ed.), *The Structure of Scientific Theories*, Second edition. Urbana, IL: University of Illinois Press, pp. 308–73.

Collins, H. M. (1981), 'Stages in the Empirical Programme of Relativism', *Social Studies of Science*, **11** (1): 3–10.

Darwin, Charles (1859), *The Origin of the Species by Means of Natural Selection*, First edition. London: Watts.

Denzau, A. T. and D. North (1994), 'Shared Mental Models: Ideologies and Institutions', *Kyklos*, **47** (1): 3–31.

Fox, M. F. (1983), 'Publication Productivity Among Scientists: A Critical Review', *Social Studies in Science*, **13** (2): 285–305.

Gibbons, Michael, Peter Scott, Helga Nowotny, Camille Limoges, Simon Schwartzmann and Martin Trow (1994), *The New Production of Knowledge: The Dynamics of Science and Research in Contemporary Societies*, London: Sage.

Hayek, Friedrich A. von (1948), 'Economics and Knowledge', in Friedrich A. von Hayek (ed.), *Individualism and Economic Order*, Chicago, IL: University of Chicago Press, pp. 33–56.

Hayek, Friedrich A. von (1973), *Law, Legislation, and Liberty, Vol. 1: Rules and Order*, Chicago, IL: University of Chicago Press.

Hodgson, Geoffrey M. (1993), *Economics and Evolution – Bringing Life Back into Economics*, Ann Arbor, MI: University of Michigan Press.

Kuhn, Thomas S. (1957), *The Copernican Revolution – Planetary Astronomy in the Development of Western Thought*, Cambridge: MA: Harvard University Press.

Kuhn, Thomas S. (1970), *The Structure of Scientific Revolutions*, Second enlarged edition, Chicago, IL: University of Chicago Press.

Langlois, Richard N. (1984), 'Internal Organization in a Dynamic Context: Some Theoretical Considerations', in Meheroo Jussawalla and Helene Ebenfield (eds), *Communication and Information Economics: New Perspectives*, Amsterdam: Elsevier, pp. 23–49.

Latour, Bruno (1987), *Science in Action: How to Follow Scientists and Engineers Through Society*, Cambridge, MA: Harvard University Press.

Polanyi, Michael (1966), *The Tacit Dimension*, Garden City, NY: Doubleday.

Psillos, Stathis (1999), *Scientific Realism – How Science Tracks Truth*, London: Routledge.

Redman, Deborah A. (1991), *Economics and the Philosophy of Science*, Oxford: Oxford University Press.

Smith, Adam ([1776] 1982), *The Wealth of Nations*, Harmondsworth: Penguin.

Stephan, P. E. (1996), 'The Economics of Science', *Journal of Economic Literature*, **34** (3): 1199–235.

Van Fraassen, Bas C. (1980), *The Scientific Image*, Oxford: Clarendon Press.

# 2. The essential tension in the social sciences: between the 'unification' and 'fragmentation' traps

**Christian Knudsen**

## 1. INTRODUCTION

The purpose of this chapter is to define which intellectual structure best promotes the advancement of knowledge within the social science disciplines. A conceptual framework will be proposed that can analyse different intellectual structures and appraise how they perform in promoting scientific progress. By the term 'intellectual structure' I refer to the distribution of activities that go on within a scientific field at a specific point in time. I will especially focus on the distribution between activities aimed at refining existing research programmes (normal science) on one hand and activities aimed at searching for new research programmes (revolutionary science) on the other hand. The question that I will explore is: What mix of the two types of activities best secures sustainable growth of knowledge in a field?

In answering this question I will put forward a framework that on the conceptual level is analogous to Schumpeter's thesis in the field of industrial organisation (cf. Schumpeter 1950). According to this thesis, neither perfect competition nor monopoly are optimal industrial structures for promoting (technological) progress. While perfect competition is too fragmented and monopoly is too concentrated, Schumpeter argued that oligopoly, by securing an optimal balance between static and dynamic efficiency, would be the market structure that best promotes (technological) progress.

By analogy with this Schumpeterian thesis, it will be argued that in order to make sustainable progress over a long period of time, a scientific field needs to secure some balance between the generation of new theoretical alternatives and the selection and retention of them. As a consequence we may find intellectual fields with either a too low or a too high degree of theoretical pluralism such that each is confronted with a specific set of problems. Fields with too little theoretical pluralism run the risk of being caught in a 'unification' trap (monopoly), while fields with too much theoretical pluralism runs the risk of being caught in a 'fragmentation' trap (perfect competition). In the first case the elaboration, modification and extension of an existing research programme tends to drive out the search for new research programmes in a self-reinforcing process. In the opposite case a field will go on searching for new programmes replacing one theory with another, without ever establishing ongoing research programmes that lead to a coherent and accumulating body of knowledge. Securing a balance between tradition and innovation therefore implies establishing a field with a few competing research programmes (oligopoly) that are constantly confronted with new tensions that drive the field towards new solutions and ultimately progress.

The chapter is organised in the following way. In the next section I describe how productive research often emerges from essential tensions either within one research programme or between two research programmes. An essential tension is defined as a problem that cannot be solved within an existing research programme but demonstrates the need for a more encompassing programme. By viewing advancement in science as a process of creative destruction it is argued, in Section 3, that it is necessary to uphold an unstable 'essential tension' equilibrium in a field in order to avoid falling into either a 'unification' trap or a 'fragmentation' trap. Different strategies for avoiding or getting out of the unification and fragmentation traps are discussed in Section 4, using economics and management studies, respectively, as main examples. Which organisational structures of intellectual fields are exposed to the two traps is discussed in Section 5, using Richard Whitley's (1984a) comparative sociology of science framework.

## 2. THE ESSENTIAL TENSION: SCIENTIFIC PROGRESS AS A PROCESS OF CREATIVE DESTRUCTION

What is productive or fruitful research? Though disagreeing on many other things, philosophers such as Kuhn (1970; 1977), Lakatos (1970), Laudan (1977), Popper (1972), and others seem to converge on the view that productive research starts from some tension, inconsistency, opposition or paradox to stimulate the development of more encompassing theories or research programmes. Since all programmes constrain a theorist's field of vision, path-breaking research will consist in removing such tensions, thereby creating new research programmes that expand the explanatory capacity of the field. Tensions, inconsistencies or oppositions may exist either within a single research programme or as an opposition between two or more research programmes in a field.

In fields composed of only one research programme, Kuhn argues that scientific revolutions would be unthinkable without the laborious puzzle-solving activity of normal science. The reason for this is that it is only through the activity of normal science that the anomalies that eventually contribute to replacement of the old paradigm by a new paradigm can be identified. There exists therefore what Kuhn calls an essential tension between tradition and innovation in science. If there is too much normal science, in the sense that normal scientists ignore anomalies by developing a 'trained incapacity' to appreciate aspects not mentioned in the paradigm, no anomalies will be taken seriously and no new research programmes will be forthcoming. If on the other hand there is too little normal science, the research community will just replace one theory with another, without establishing any ongoing research programmes. Or as Kuhn (1970, p. 65) argues:

> By ensuring that the paradigm will not be too easily surrendered, resistance guarantees that scientists will not be lightly distracted and the anomalies that lead to paradigm change will penetrate existing knowledge to the core. The very fact that a significant scientific novelty so often emerges simultaneously from several laboratories is an index both to the strongly traditional nature of normal science and to the completeness with which that traditional pursuit prepares the way for its own change.

While researchers try to solve a problem within a research programme they may end up creating solutions that destroy their original paradigm. Such a pattern of (scientific) development, where a research programme contains the seeds of its own destruction, closely resembles the 'creative destruction processes' studied by Schumpeter in his work on the transformation of capitalist economies.

Hayek (1948) has, for instance, argued that the 'knowledge problem' in economics – that is, the problem of explaining how perfect rational agents can obtain enough knowledge about each other's actions in order to reach an equilibrium solution – may be such an 'essential tension' or paradox in the maximisation paradigm. The paradoxical nature of this problem is caused by the fact that any solution to this problem necessarily leads to a destruction of the maximisation programme, because we need to introduce some kind of evolutionary mechanism working behind the backs of the agents; in other words, the question of how equilibrium comes about cannot be posed in fully orthodox theoretical terms because it leads to a self-reference problem that makes it impossible to explain how economic agents in an interdependent system can acquire perfect knowledge about each other's decisions. This implies that trying to solve this problem within the maximisation framework destroys the framework, pointing in the direction of an evolutionary-institutional research programme (cf. Knudsen 1993).

While Kuhn, with his predominantly mono-paradigmatic view of science, focused mainly on tensions or paradoxes within a single research programme, other philosophers of science have directed our attention to multi-paradigmatic situations, where there exist tensions or oppositions between different research programmes in a field. However, just as in the mono-paradigmatic case above, we may also have too much tradition or too much innovation, depending upon how the community of researchers decides to deal with tensions and oppositions between theories (Holmwood and Stewart 1994; Poole and Van de Ven 1989). If the researchers in a field with several research programmes follow an isolationist strategy by continuing the internal development of their own research programme ignoring what goes on in other research programmes, the researchers choose to live with the tensions and oppositions instead of seeing them as opportunities for creating new ways of looking at the world that may lead to new synthesis and new insights. However, we may also commit the opposite mistake by thinking that these oppositions or contradictions between research programmes may be resolved very easily by just testing the theories against the same empirical data in order to find out which one fits the data best. As Imre Lakatos (1970) reminded us, there is no such thing as a 'crucial test' that will immediately decide between two competing theories or research programmes.

Neither of the two strategies above seems, however, to uphold an essential tension that can drive the field through a creative destruction process by extending the boundaries of existing theories, building more encompassing theories or syntheses. Or as stated by Van de Ven and Poole (1988, p. 25) in relationship to organisation and management theory

> There are many ringing denunciations of opposing viewpoints, but too few attempts at bridging or synthesis. Hence, addressing organizational paradoxes is an exciting and challenging effort. It is an issue on the cutting edge of organization and management theory, and one that will spawn new ideas and creative theory. Looking at paradoxes forces us to ask very different questions and to come up with answers that stretch the boundaries of current theories.

A very similar idea of an essential tension has been put forward by the economist Paul Krugman (1996, p. 140), known for the modern trade theory:

> I have not encountered a single example of a great innovator who was not immersed in his or her field, deeply familiar with its tradition – and therefore able to be truly creative in challenging that tradition. So here, finally, is the sermon. By

all means, try to be a radical innovator – a crazy economist. But understand that being crazy in a productive way is hard work. And realize, in particular, that you are very unlikely to have interesting new ideas unless you have learned to appreciate and respect interesting old ideas.

This Kuhnian view of productive science as approaching an essential tension between tradition and innovation, initiating processes of creative destruction, is in accordance with the 'correspondence principle' of the Danish physicist Niels Bohr. Popper (1972, p. 202) gives the following explanation of this principle:

I suggest that whenever in the empirical sciences a new theory of a higher level of universality successfully explains some older theory by correcting it, then this is a sure sign that the new theory has penetrated deeper than the older ones. The demand that a new theory should contain the old one approximately, for appropriate values of the parameters of the new theory, may be called (following Bohr) the 'principle of correspondence'.

A 'correspondence view' of scientific advance can be interpreted as dialectic in the sense that all problems in research emerge from tensions, contradictions or oppositions either within a single research programme or between two or more research programmes.

According to the correspondence principle, tensions, contradictions or paradoxes may emerge because the conceptual framework used in a theory T1 is too narrow to understand a specific phenomenon. In order to remove this tension or contradiction in theory T1, we may try to broaden the conceptual framework of the old theory to include what was excluded before, thereby constructing a more general theory T2. A scientific advance will therefore consist in the establishment of a correspondence relationship between the two theories. Such a relationship will exist if T1 is a satisfactory approximation for T2 within the domain D1, but T2 corrects the explanations/predictions of T1 outside that range, that is, in D2–D1.

Proponents of an evolutionary research programme in economics such as Nelson and Winter (1982), Schumpeter (1950), and Winter (1975) have all used the 'correspondence view' of scientific advance as an argument for replacing the standard neoclassical research programme with an evolutionary theory. Like Hayek's 'knowledge problem', the dilemma or paradox that confronts the neoclassical research programme and leads to the necessity of constructing a broader evolutionary research programme, has been described in the following way by Nelson and Winter (1982, p. 27):

Thoroughgoing commitment to maximization and equilibrium analysis puts fundamental obstacles in the way of any realistic modelling of economic adjustment. Either the commitment to maximization is qualified in the attempt to explain how equilibrium arises from disequilibrium or else the theoretical possibility of disequilibrium behaviour is dispatched by some extreme affront to realism.

Since the neoclassical programme cannot solve the adjustment problem, this problem then leads, according to Winter (1975, p. 96), to an evolutionary research programme, the purpose of which is '... to develop a more fundamental theory that explains both the range of validity of the approximations and the phenomena that lie outside that range'. In accordance with the 'correspondence view' of theoretical advance, the evolutionary theory is used to determine the limited domain D1 of the neoclassical theory as well as studying new phenomena outside this range D2–D1:

> The qualitative predictions of orthodox comparative static may well describe the typical pattern of firm and industry response in the dynamic, evolving economy of reality. However, evolutionary analysis probes more deeply into the explanations for these patterns and warns of possible exceptions. Also the explicit recognition of the search and selection component of adjustment brings a whole new range of phenomena into theoretical view. (Nelson and Winter, 1982, p. 175)

## 3. BETWEEN THE 'UNIFICATION' AND THE 'FRAGMENTATION' TRAPS

Building upon the discussion in Section 2, intellectual fields often experience problems of securing an essential tension between tradition and innovation. But what is worse, besides having problems with securing such a balance, intellectual fields are also constantly exposed to traps that may drive them into either a self-reinforcing spiral of elaborating upon existing programmes or into a self-reinforcing spiral of search for new research programmes.

In both cases, the possibilities of keeping an optimal balance between extending existing research programmes versus searching for new programmes or shortly securing an unstable 'essential tension' equilibrium will be upset. In the following section I will try to explicate the mechanism driving each of these traps as well as discussing how these two traps interacts to upset any balance between them.

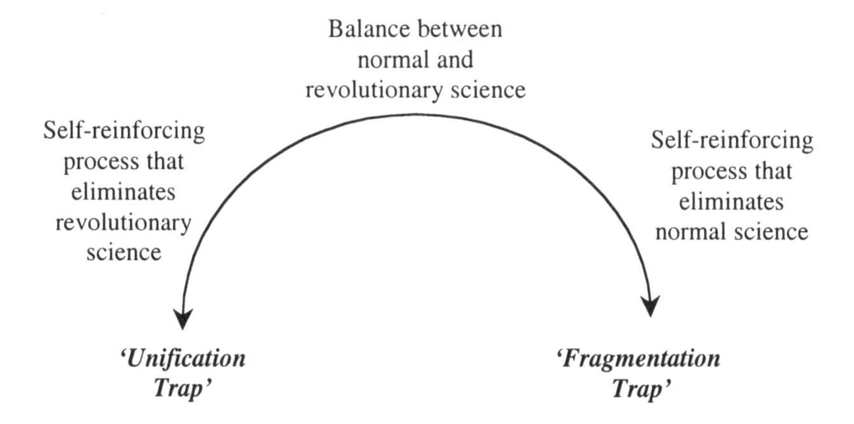


*Figure 2.1 The 'unification' and 'fragmentation' traps*

**3.1 The 'Unification' Trap and the Case of Economics in the post-World War II Period**

The 'unification' trap is present when normal science drives out revolutionary science and the activity of elaborating, modifying and extending an existing research programme gradually comes to dominate the search for new research programmes through a self-reinforcing process. As researchers in a field develop better and better skills in using the problem-solving heuristic of an existing research programme, this programme will be even more used to solve new problems, thus further increasing the strength of the positive heuristic and the opportunity costs of searching for new research programmes. Acquiring competencies to solve problems within one research programme therefore increases specialisation within that programme, making it more and more difficult for alternative research programmes to compete. This self-reinforcing process may lead to a 'unification' trap where all research activity in the field goes on within a single research programme instead of being distributed between several competing research programmes. The 'unification' trap therefore emerges because the exploitation of already-existing research programmes gives a faster and safer return than does experimentation with new and uncertain research programmes. The 'unification' trap consequently implies a scarcity of exploratory activities that in the long run undermines the flexibility of the field by reducing its ability to adapt to new and unpredictable situations.

Within the social sciences, economics is probably the only field that has been caught in a 'unification' trap for an extended period of time during the hegemony of the maximisation paradigm after World War II. With the development of this programme, economists developed a more and more refined mathematical heuristic that made it more and more attractive to use the neoclassical research programme and its positive heuristic, and less and less attractive to switch to any alternative programme's heuristic. This self-reinforcing process of the 'unification' trap led, however, to an imbalance where heuristic progress (that is, the development of the positive heuristic/problem-solving methodology) came to dominate the empirical problem-solving activity in the field. For economists who believe in efficient markets such a conclusion seems rather unlikely. Or as Grubel and Boland (1986, p. 421) state:

> Economic knowledge and human capital are sold in markets. For most economists this implies a strong presumption that both are priced correctly and produced efficiently. Any university using too few or too many mathematics-teaching economists should find that its graduates are at a competitive disadvantage; its training program should shrink and finally disappear. Similarly, knowledge that contains inappropriate amounts of mathematics should lose out in the market and its production will contract or cease.

But how is it possible that mathematical modelling of economic phenomena within the neoclassical research programme during the post-World War II period became the high prestige area in economics while applied research gave a much lower return in terms of reputation? According to Grubel and Boland (1986), this may be explained using a model of rent-seeking behaviour. In such a model it is assumed that there are two groups of economic theorists: mathematical and applied researchers, which both attempt to generate economic rents for their members. The relatively lower rate of return to the rent-seeking behaviour of applied researchers is explained in the following way. First, since mathematical economists have fewer employment opportunities outside academia than applied economists, mathematical economists as a group will have greater incentives to seek rents in the university environment. Second, since it is easier to formulate objective tests of competence in mathematical than within applied economics, mathematical researchers will find it easier to build barriers to entry and therefore to defend rents than any other group of researchers. Third, since mathematical economists are not facing any direct market tests like applied researchers, the demand for their products can be stimulated by themselves, leading to higher rents and reputation than other groups of researchers. Fourth, due to the universality of mathematics as a language, mathematical economists can build coalitions with other natural scientists, statisticians, econometricians and of course, mathematicians, thereby achieving a higher degree of internationalisation than other subgroups in economics. The result is that mathematical economists through network externalities will earn higher rents and reputation than other groups of researchers. Fifth, this dynamic process of reinforcement of mathematical economics was initiated in the early 1950s and 1960s. It received stimulus both from great optimism about the usefulness of natural science methods to the social sciences and government support for the training of mathematical economists in the *post-Sputnik era*.

While economics was caught in a 'unification' trap with the elaboration and extension of the positive heuristic of the neoclassical research programme during the 1950s, 1960s and 1970s, the field seems to have escaped the trap during the latter part of the 20th century. While – with the exception of the old institutional economics – there were few heterodox traditions in economics in the period after World War II, things started to change around the later part of the 1970s and the 1980s. At that time a whole set of theories – many with roots in the pre-World War II period and earlier – were marketed as new heterodox research traditions in the economic profession. Among those were transaction cost economics, the evolutionary research programme, Austrian economics, post-Keynesian tradition, property rights economics, information economics, and so on. This signalled that economics had managed to get out of the 'unification' trap and that heterodox traditions actually managed to influence the type of problems taken up by mainstream economics.

### 3.2 The 'Fragmentation' Trap and the Case of Management and Organisation Studies

The second trap is called a 'fragmentation' trap and is present when revolutionary science drives out normal science and the search for new research programmes comes to dominate the elaboration, modification and extension of existing research programmes. There are several reasons why a scientific field may end up in a 'fragmentation' trap. First, most new scientific ideas will be worse than the existing pool of ideas. Second, it takes a lot of time and experience before the positive heuristic of a new research programme can be developed enough so that normal scientists can successfully exploit it. Even the research programme that turns out to be most successful will normally perform rather badly to start with. Due to a lack of persistence in the scientific community many theories may therefore never be investigated well enough to become programmes for research, before new theories have been proposed and have replaced them. The real potential of a theory to become a new research programme will therefore never be discovered. And when the process – which drives new theories to be introduced in a field without replacing older theories – takes on a self-reinforcing character, the field ends up in a 'fragmentation' trap with no chain of coherence through time or accumulation of knowledge.

In some cases the 'fragmentation' trap is due to a 'fad and fashion' mentality that implies that new approaches are introduced into a field at a faster and faster speed. In management studies Harold Koontz (1961) talked early on about 'the management theory jungle'. Nineteen years later he concluded, 'the jungle appears to have become even more dense and impenetrable' (Koontz 1980: 175). In *Organization Studies*, Lex Donaldson (1995, pp. 7-8) argues, 'since around 1967 at least fifteen new paradigms have been launched ... on average a new paradigm is offered every second year'. Such a process of proliferation will typically start when a 'value of novelty' (Pfeffer 1993) or a 'uniqueness value' (Mone and McKinley 1993) that favours new ideas rather than integration and consolidation becomes dominant in a field. According to Mone and McKinley (1993) such a value has emerged within the fields of organisation and management studies, as documented from statements of leading authorities, and editors of leading journals such as *Administrative Science Quarterly*, *Academy of Management Review* and *Organization Science*. This value 'prescribes that uniqueness is good and that organisation scientists should attempt to make unique contributions to their discipline' (Mone and McKinley 1993: 284). Similarly, Pfeffer (1993, p. 612) states that: 'Journal editors and reviewers seem to seek novelty, and there are great rewards for coining a new term. The various divisions of the Academy of Management often give awards for formulating "new concepts" but not for studying or rejecting concepts that are already invented'.

However, the introduction of new approaches in both management and organisation studies – due to the 'uniqueness value' – leads according to Mone and McKinley (1993) to a problem of 'information overload'. The more and the faster new paradigms are introduced into organisation and management studies, the less intellectual capacity will be available for exploiting and appraising the existing paradigms. This implies that paradigm proliferation will shift resources from normal to revolutionary science in a self-reinforcing manner. This leads to a 'fragmentation' trap as formulated by Donaldson (1995: 10):

> With the constant rush to the next paradigm the consequence is half-finished research programmes, as exemplified by structural contingency theory, where decades of research have left a literature widely perceived as containing unresolved theoretical problems and empirical inconsistencies ... Reference to such problems is a standard argument for embarking upon the next new paradigm, but this argument can be self-defeating, precluding the completion of any research programme.

As argued by Zammuto and Connolly (1984) and Van de Ven (1999) the problem of 'information overload' – while being caught up in a 'fragmentation' trap – may be especially problematic to handle for new doctoral students. They will be confronted with a bewildering diversity of theories that they will have no chance of digesting. The background knowledge of the field will be 'a morass of bubbling and sometime noxious literature. Theories presented are incompatible, research findings inconsistent, and the general body of knowledge indigestible' (Zammuto and Connolly, 1984: 32). The combination of the 'uniqueness value' and a very fragmented knowledge structure will make it tempting for many doctoral students to learn only some of the newer and more exciting programmes at the expense of the old. In this way, the field may end up in a situation where there is no effective transmission of knowledge between the old and the new generation and where the older theories are not replaced by newer theories, but just forgotten. Being caught in a 'fragmentation' trap implies therefore that the historical dimension of a field will tend to get lost. Few attempts will be made to show how newer contributions relate to earlier contributions by expanding upon and correcting older contributions. Instead newer contributions will just be introduced into the field without facing any demand that they somehow should solve problems that earlier contributions had been unable to solve.

### 3.3 How Tradition and Innovation Interact to Undermine a Healthy Balance Between Themselves

In order to secure the unstable 'essential tension' equilibrium, scientific fields may be seen as constantly trying to avoid getting locked into either a self-reinforcing 'unification' or a 'fragmentation' trap. However, there exist very complex interactions between activities of exploiting an existing research programme and activities of searching for new theories that will tend to undermine any kind of balance that may exist between them.

Elaborating, modifying and extending an existing research programme tends to undermine extraordinary science by discouraging attempts at finding new research programmes and problem-solving heuristics that are essential for the long-term survival of a field. Researchers in the field therefore either tend to stick to one (currently progressive) programme and its problem-solving heuristic to such an extent that there is little exploration of other programmes. Or they fail to stick to one (underdeveloped and currently degenerating) programme long enough to determine its 'true' problem-solving capacity.

In a similar way, revolutionary science undermines normal science. Efforts to promote revolutionary science encourage impatience with new theories and make the development of new problem-solving heuristics very unlikely. Theories are therefore likely to be abandoned before enough time has been devoted to developing them into research programmes with a specific heuristic. The impatience of revolutionary science therefore results in unelaborated discoveries and a fragmented knowledge structure. As a result of the way normal and revolutionary science interact, most scientific fields will have difficulties maintaining a healthy tension between them. This tendency to undermine each other raises the problem of what strategies scientific fields have in fact used in order to keep a balance between them, thereby avoiding both the 'unification' and 'fragmentation' traps. In other words what kind of 'competition policies' have been implemented in different scientific fields?

## 4. STRATEGIES FOR AVOIDING THE 'UNIFICATION' TRAP AND THE 'FRAGMENTATION' TRAP IN THE SOCIAL SCIENCES

By arguing that self-reinforcing processes and traps characterise intellectual fields several important policy issues may be brought forward. For instance, if positive feedback loops might lock in a dominant research programme such as neoclassical economics for long periods of time, regardless of the intellectual progress actually generated by it, is it then sensible to talk about an open market for ideas? What can be done about the substantial sunk costs and high entry barriers that confront unorthodox competitors? Conversely, what policies may be recommended in case a field is exposed to a feedback loop that leads to its fragmentation? When one theory after the other is introduced in rapid succession, it will be impossible to evaluate which ones are good and which ones are bad, thereby destroying the possibility of having an accumulating body of knowledge. In this case, we may investigate policies to reduce the speed with which new alternatives are introduced into the field.

Before entering into discussions about different strategies to avoid dilemmas and traps in a scientific context, I will offer a few reflections upon how such discussions should be conducted. First, I will suggest that debates about appropriate strategies should be conducted in a comparative context so that we avoid policy conclusions that rest only upon experiences from a single field, but use data from a broader set of fields. In this way we may avoid the 'Panglossian bias' of many researchers who view the structure of their own field as the only natural/possible way to organise a field. Second, policy debates should preferably be conducted on a constitutional level, which implies that the discussions about what rules (conventions, norms, and so on) should govern a specific field preferably should be conducted behind a 'veil of ignorance' (Rawls 1971) or a 'veil of uncertainty' (Buchanan and Tullock 1962); that is, without knowing what implications these rules may have regarding the choice between specific theories or research programmes.

## 4.1 The 'Unification' Trap and the Case of Economics

In the following section we shall discuss some strategies that a research community may suggest in order to avoid a self-reinforcing 'unification' trap, using economics as the main illustrative case. These strategies include (a) promoting the isolation of young heterodox research programmes during their maturation; (b) building heterodox traditions around core anomalies in mainstream economics and giving priority to a strengthening of their positive heuristic: and (c) changing the composition of research styles in an intellectual field.

### 4.1.1 Promoting the isolation of young heterodox research programmes during their maturation

In his *The Open Society and Its Enemies* Karl Popper (1945) argued that the more 'open' a scientific field is in terms of accepting competing research programmes, the tougher the competition and the better the chances of a scientific breakthrough. For the same reason, the scientific community should be very lenient towards new research programmes, in order to make sure that they get enough time to mature, before being exposed to the fierce competition of older and more mature research programmes. In accordance with this position, Imre Lakatos (1970, p. 157) argued that 'We must not discard a budding research programme simply because it has so far failed to overtake a powerful rival. As long as a budding research programme can be rationally reconstructed as a progressive problem shift, it should be sheltered for a while from a powerful established rival'.

In economics, for instance, the maximisation programme has been elaborated, modified and extended for many years. A lot of sunk costs have been spent on this programme and it is, therefore, very unlikely that a competitor with a better problem-solving capacity or a stronger heuristic will suddenly emerge. New research programmes such as the behavioural programme, the evolutionary programme, the new institutional programme and so on, should therefore be protected during their infancy in order to make sure that they get time to develop and strengthen their heuristic before a verdict can be made. In fact, this argument is analogous to the 'infant industry' argument, recommending that new firms should be protected from outside competitors until they have grown strong enough to be exposed to the fierce competition of the world market from older and more mature foreign competitors.

Terence Ball has raised a similar argument in political science. He argues that Marxism and functional analysis as scientific programmes might have been killed off prematurely, because their protagonists lacked the necessary tenacity and their critics the necessary tolerance required to give these programmes a fighting chance. He concluded (Ball, p. 172):

> We political scientists have not, I fear, treated our budding research programs leniently. We have, on the contrary, made them into sitting ducks; and, in a discipline, which includes many accomplished duck hunters, this has often proved fatal ... If we are to be good sportsmen we need to take Lakatos' methodology seriously.

This implies that we should be more tolerant when criticising a research programme than is assumed by Popper's methodology of naïve falsificationism. Purely negative and destructive criticism, like a refutation, will never be enough to kill a programme. We will only be able to reject a programme when we have a new and better theory or programme. This leads us to the second strategy.

#### 4.1.2 Building heterodox traditions around core anomalies in mainstream economics and giving priority to a strengthening of their positive heuristic

Though economics has often been portrayed (mostly by other social sciences) as having a completely unitary structure, today the field includes several heterodox traditions that during the last 20 or 30 years have influenced mainstream economics in fundamental ways. In comparison with the older heterodox traditions such as the old institutionalism, some of the new heterodox traditions have not been marginalised in the same way. Consequently the field has recently been able to move out of the 'fragmentation' trap, getting closer to an 'essential tension' equilibrium with a better mix of normal and revolutionary science. The main reason is that the new heterodox traditions seem to have followed a strategy that on the one hand focuses on the core anomalies in the mainstream paradigm and on the other hand gives priority to strengthening the positive heuristic of these newer research programmes.

New heterodox traditions such as transaction cost economics, the evolutionary research programme and the knowledge-based programme start from the assertion that a purely negative critique will be insufficient in order to replace the mainstream tradition. What is needed is that the new heterodox traditions are able to identify some core anomalies in the orthodoxy and to show how a solution to these problems will lead to the replacement of the mainstream tradition with a newer research programme. An example of such a core anomaly is the so-called 'knowledge problem' that potentially may be solved by switching from the mainstream tradition to an institutional-evolutionary research programme. The argument is that you will never 'beat' a research programme just by identifying some anomalies within it. In order to do so you need a new and better research programme. It is for the same reason that several new heterodox traditions have allocated so many resources to improving the problem-solving capacity of the new research programmes by strengthening their positive heuristics.

#### 4.1.3 Changing the composition of research styles in an intellectual field

When studying scientific fields most philosophers of science tend to view science from a typological rather than from a population perspective (cf. Mayr 1976). A typological view of a scientific field assumes that all researchers within a field follow the same rule of rationality, converging on the same uniform decision, and which therefore approximates the same common underlying ideal type. Consequently, variety or the co-existence of a plurality of approaches will either be very difficult or even impossible to understand from such a perspective. For instance, in the view of Imre Lakatos Methodology of Scientific Research Programmes, continuing the work on an old degenerating research programme when a new progressive programme emerges can only be understood as involving a 'mistake' or 'irrational behaviour' on the part of the researchers staying with the old programme.

Variety or the co-existence of several research programmes takes centre-stage in a population perspective. In this case, we switch to a truly system (population) level of analysis by asking what distribution of individual research strategies may be rational (conducive to scientific progress) for the field as a whole (cf. Kitcher 1993). Let us assume that the research community may be described by different styles of research. First, there is the 'orthodox normal scientist' who will stick with the paradigm, whatever happens. Second, there is the 'standard normal scientist' who will be ready to reject an old paradigm when there are clear indications that a new paradigm will supersede the old paradigm. Third, there is the 'essential tension researcher' who exploits anomalies in the old paradigm(s) in order to create new research programmes. And finally, there is the 'fashion-driven' researcher who values new research programmes more than old research programmes just because they are newer.

The existence of an uneven distribution of research styles in a field, with a dominance of the first-mentioned, will lead to a 'unification' trap. In this case, a strategy for getting out of this trap would consist in trying to create a more even distribution by favouring the last-mentioned research styles when recruiting to the field. This may be done by consciously promoting entry of new heterodox researchers into the field.

## 4.2 The 'Fragmentation' Trap and the Case of Management/-Organisation Studies

In the following section we shall discuss three strategies that researchers may use in order to avoid a self-reinforcing 'fragmentation' trap. The illustrations will come from the field of management/organisation studies. These strategies include (a) increasing the persistency within the research community; (b) focusing on removing tensions and oppositions between theories and research programmes; and (c) condensing the knowledge structure in order to increase the absorptive capacity of an intellectual field.

### 4.2.1 Increasing the persistency within the research community

An important reason why a field may end up in a 'fragmentation' trap is that there is too little persistence and too much impatience in the scientific community. Since new theories and approaches need a lot of refinement and development before their true value in terms of heuristic power (problem-solving capacity) can be determined, the high rate of 'turnover' will imply that the selection mechanism will function very imperfectly. As I argued earlier the individual researchers as well as the research community will be confronted with a problem of 'information overload' (Mone and McKinley 1993), since new theories are introduced into the field at a rate that makes it impossible to identify the theories with the best heuristic or problem-solving capacity. In other words the variation mechanism that produces new theories will completely outperform the selection mechanism. As a consequence, theories will just succeed each other in an endless cycle of failure and change without any real accumulation of knowledge. That is, the field is in a typical 'fragmentation' trap.

Fields that ends up in a 'fragmentation' trap rarely have a good sense of their own history because the researchers are too busy keeping up with the newest 'fad and fashion' to look back upon the historical roots of their field. Or as Reed (1992, p. 246) has argued in relationship to the field of organisational studies:

> Any sense of historical continuity and narrative coherence is lost in the clamour of voices announcing the 'end of history' and extolling the virtues of root and branch transformation from the 'old' to the 'new' organization theory.

Furthermore, as we shall discuss below, the knowledge structure is too fragmented and diffuse to be of any help to researchers in their current problem-solving activities. Introducing a 'conservative bias' towards the status quo in such fields would therefore be a strategy that can reduce the high 'turnover' rate of new theories, thereby securing a better balance between continuity and change. One way that such a 'conservative bias' strategy could be implemented is to demand that only theories that build upon and correct older theories should be taken as serious candidates for being included in the knowledge structure of the field.

#### 4.2.2 Focusing on removing tensions and oppositions between theories and research programmes in the field

In cases where a field is caught in a 'fragmentation' trap any attempt to isolate theories or research programmes rather than confront them with each other will be seen as counterproductive. For instance, both in management studies and organisation studies there are many supporters of an eclectic research strategy (cf. Mintzberg, Ahlstrand and Lampel 1998). However, supporters of this strategy often seem unaware of the negative implications that such a strategy may have regarding the knowledge structure of a field by moving the field away from an 'essential tension' equilibrium.

An eclectic research strategy will often lead to a high degree of theoretical conservatism rather than theoretical radicalism. The reason for this is that adherents of the eclectic strategy are willing to accept and live with very varied – and in some cases even contradictory – theoretical perspectives. This implies that inconsistencies and tensions in the field are accepted, rather than acted upon. Consequently, tensions and inconsistencies in the knowledge structure of the field are not seen as leading to new research problems and new research opportunities, though a closer investigation of them might have revealed that they could have done so. Another implication of the eclectic type of research strategy is that the members of the research community will find it more and more difficult to communicate with each other the more theories that are introduced and the further they get stuck in the 'fragmentation' trap.

The only way to reverse this tendency towards a 'fragmentation' trap is therefore to give up the eclectic research strategy and instead focus on removing tensions and opposition between theories and research programmes. This necessitates continuous investments in research that tries to solve all kinds of conceptual problems that emerge as contradictions or tensions between the different parts of the knowledge structure. Since these conceptual problems (and the kind of foundational research that is implied) are often looked upon as being of a second-order, they are not considered as important to solve as empirical problems.

#### 4.2.3 Condensing the knowledge structure in order to increase the absorptive capacity of an intellectual field

A continuous accumulation of new theories and methods in a field will sooner or later create a complexity crisis, because the knowledge structure will become too diffuse and disintegrated. Adding new layers of knowledge in a field, without at the same time trying to 'condense' the knowledge structure by ordering/mapping the different theoretical contributions, will sooner or later lead to severe inefficiencies that reduce the absorptive capacity of an intellectual field. Or as Jeffrey Pfeffer (1982, p. 1) states: 'There are thousands of flowers blooming, but nobody does any manicuring or tending'. In a field that is exposed to such a 'fragmentation' trap the researchers will experience difficulties in 'standing on the shoulders of their predecessors', because there is no real structure of knowledge to start from.

One way to reverse this self-reinforcing tendency toward a 'fragmentation' trap is to make investments in structuring the background knowledge of the field. That is, changing the balance between continuity and change by moving resources from the latter to the former. This may be done, for instance, by using resources to map the existing theoretical contributions within a field, by identifying the main dimensions along which they differ, by specifying the formal and substantial relationships between individual contributions, by identifying important oppositions between different contributions, and so forth. In organisation and management studies, examples of such contributions are Astley and Van de Ven (1983), Burrell and Morgan (1979), Pfeffer (1982), Scott (1998) and others. Structuring the background knowledge of a field in this way will make it more likely to be used in future problem-solving activities. It will therefore be a strategy for increasing the absorptive capacity of a field and it will also increase the chances of finding more encompassing theories that may further reduce the fragmentation of a field.

Working on 'condensing' the knowledge structure in a field may have significant efficiency implications, since this determines to a large degree the production time for new knowledge contributions. If the knowledge structure is very complex, consisting of many layers of old knowledge that has only been condensed to some degree, the production time of new contributions is relatively high. One direct way to measure or estimate this 'production time' is to find out how much time a newcomer to the field needs to recapitulate the history of the field, in order for him or her to make a new knowledge contribution. The idea that a new member of a scientific field 'recapitulates' in a condensed version the evolution of the field goes back to the thesis that 'ontogeny recapitulates phylogeny' in biology. This thesis states that every new member of a species recapitulates in a condensed way the evolution of the whole species. In a scientific context, according to Herbert Simon (1962), this thesis states that every new member of a scientific field will recapitulate – in a very condensed way – the historical evolution of the field, in order to make a new knowledge contribution.

In fields where a lot of resources have been spent on formalising and thereby condensing the knowledge structure, the production time and the age of new contributors will be lower than in fields where this is not the case. However, investments in removing tensions and oppositions, thereby keeping the knowledge structure relatively simple and manageable are primarily done in order to increase the absorptive capacity of the researchers within the field (cf. Cohen and Levinthal 1990); that is, to secure long-term scientific progress in the field.

# 5. THE 'UNIFICATION' VERSUS 'FRAGMENTATION' TRAP IN 'POLYCENTRIC OLIGARCHIES', 'PARTITIONED BUREAUCRACIES' AND 'FRAGMENTED ADHOCRACIES'

In the following section I will show how the discussion of the 'unification' trap versus the 'fragmentation' trap fits into Richard Whitley's (1984a) discussion of the intellectual organisation of scientific fields. Whitley argues that it is possible to identify very different modes of how scientific fields are organised as reputational systems based on two dimensions: 1) the degree of interdependency and 2) the degree of task uncertainty.

The degree of interdependency refers to what degree researchers in a field are dependent on each other to obtain reputation. The more applied a field is, the more open it will be towards its environment and the less interdependency there will be. Conversely, the more basic a science is, the more researchers have to rely on each other for obtaining reputation. The degree of task uncertainty refers to the degree of uncertainty a researcher faces when trying to solve a specific problem. It is normally claimed that the main function of science is to produce new knowledge. What is accepted as new knowledge depends to a large extent on the background knowledge of the field. As I argued in Section 4, the more systematic, exact and general this knowledge is, the easier it is to determine whether a contribution is new or not and how well this contribution fits into the background knowledge of the field. If the background knowledge is well-structured, which is the case for mono-paradigmatic fields, the task uncertainty will be low.

Whitley (1984a) furthermore distinguishes between two different aspects of task uncertainties, technical and strategic. Technical uncertainty refers to the degrees of unpredictability and variability that exist in a field with regard to the methods and procedures for solving empirical problems. If many different methods exist and if it is difficult to interpret the (test) results in a field, the degree of technical uncertainty is high. Opposite, if a certain method has been canonised as being the only legitimate method in a field, the degree of technical task uncertainty is low. Strategic uncertainty, on the other hand, refers to the degree to which researchers agree upon which problems are important, less important, and so on, and what goals should govern their research. In fields with a high degree of strategic task uncertainty, researchers will be confronted with many different problems, the relevance and importance of which are appraised very differently.

According to Whitley, variations in these two contingency variables make it possible to distinguish between at least seven different configurations of how scientific fields are organised. However, since we are primarily interested in the social sciences, the discussion here will be limited to the three configurations found in this area. These include the 'fragmented adhocracy', the 'polycentric oligarchy' and the 'partitioned bureaucracy'.

*Table 2.1 Reputational organisations in the social sciences*

| **Degree of strategic task uncertainty** | **Degree of interdependency** | |
|---|---|---|
| | **Low** | **High** |
| High | 'Fragmented adhocracy' | 'Polycentric oligarchy' |
| Low | 'Unstable form' | 'Partitioned bureaucracy' |

Social science fields such as sociology, management studies, anthropology, political science and so on have rarely been dominated by a single paradigm, as is the case for some natural sciences. We should therefore expect that these fields have a substantially higher degree of technical and strategic uncertainty. The only social science that has diverged from this pattern is economics, which for a long period has been dominated by the (neoclassical) maximisation paradigm and therefore has a substantially lower degree of strategic task uncertainty and a higher degree of interdependency than the other social sciences. According to Richard Whitley (1983), the reputational configuration of economics may be characterised as 'partitioned bureaucracy'.

As a 'partitioned bureaucracy', economics consists of a core with pure and abstract theorising (within the maximisation paradigm) and a number of peripheral sub-fields of applied research. Due to the absence of control over the object of research and the ambiguity of empirical testing in the social sciences, any unifying theoretical framework in a social science will be under permanent threat of being replaced. In economics, however, this problem was solved by partitioning the core of pure theory with formal mathematical modelling from applied and empirical research in the peripheral areas. Compared to other ways of organising social science fields, economics has a very hierarchical type of reputational organisation, since research in the core of the field is viewed as much more prestigious than applied research in the sub-fields. The term 'partitioned' in 'partitioned bureaucracy' refers to the absence of feedback from the applied research in the periphery to the pure theory in the core, in other words, the abstract models of the maximisation paradigm have been 'immunised' from 'potential falsifications' arising in the applied field.

The second type of organisational configuration found in the social sciences according to Whitley (1984a) is the 'polycentric oligarchy'. Examples of this structure are classical Continental sociology, British social anthropology and, as I have argued in a recent paper, organisation theory in the US after 1975 (cf. Knudsen 2003). The 'polycentric oligarchy' emerges when relatively small groups of researchers gain control over critical resources such as positions and journal access. But since the degree of task uncertainty is very high, their control can only be exercised locally and personally, resulting in the establishment of several independent centres. In organisation theory these centres were formed around the main research programmes that emerged in the late 1970s such as population ecology, transaction cost economics, institutional theory and resource-dependency theory. Within each of the centres, there was a relatively strong hierarchical reputational organisation due to a consensus on what was the basic framework to be used, what were the important problems to be solved and therefore how reputation should be allocated within the 'specialised' research community. However, there was very little coordination and cooperation between the centres, and intense competition in order to gain control over the whole field. Consequently the field became balkanised into a set of more or less autonomous centres, each pursuing their own research agenda, with minimal interaction and communication.

The 'fragmented adhocracy' that may be found in management studies (Whitley 1984b) is characterised by a low degree of interdependency between researchers, which implies a rather loose research organisation. Since researchers are facing few restrictions regarding the choice of theoretical framework and choice of method, the degree of technical and strategic task uncertainty is very high. This breeds a fragmented knowledge structure and disagreement about the relative importance of the problems to be solved by the field. As a result, the problem-solving activity within the field takes place in a rather arbitrary and ad hoc manner, with limited attempts to integrate new solutions with the existing structure of knowledge.

But how does the framework presented in this chapter fit into Whitley's comparative analysis of intellectual fields? There appears to exist a very simple answer to this question following from the discussion above. If a field is very hierarchical in its reputational organisation, which is the case for the 'partitioned bureaucracy' (low degree of strategic task uncertainty and a high degree of interdependency), the field will typically be struggling to avoid or get out of a 'unification' trap. If the field, on the other hand, has a very flat reputational configuration, which is the case for 'fragmented adhocracies' (high degree of task uncertainty and low degree of interdependency), the field will be struggling to avoid or get out of a 'fragmentation' trap. Fields with an organisational structure that is situated between these two extremes, such as the 'polycentric oligarchy', will on the other hand be closer to maintaining an unstable 'essential tension' equilibrium between tradition and innovation or a balance between elaborating existing research programmes and searching for new programmes.

## 6. CONCLUSION

Philosophers of science such as Karl Popper (1945) have argued that the more diverse or pluralistic a field becomes the tougher the competition will be and the better will the chances of a scientific breakthrough be. In this chapter I have argued that such a policy prescription between the degree of pluralism and scientific progress need not be generally valid across all fields. In accordance with the Schumpeterian thesis in competition policy, more pluralism may have positive as well as negative consequences for scientific advance in a field, depending upon how far the field is from the unstable 'essential tension' equilibrium. If a field is already caught in a 'fragmentation' trap a policy of pluralism will be counterproductive, since it will just lead to more and not less fragmentation. If the field, on the other hand, has been caught in a 'unification' trap a policy of increasing theoretical pluralism may have a positive effect. The structure of a scientific field that best seems to avoid both the 'fragmentation' trap and the 'unification' trap is the 'polycentric oligarchy'.

## REFERENCES

Astley, W. G. and A. H. Van de Ven (1983), 'Central Perspectives and Debates in Organization Theory', *Administrative Science Quarterly*, **28** (2): 245–73.

Ball, T. (1976), 'From Paradigms to Research Programs: Towards a post-Kuhnian Political Science', *American Journal of Political Science*, **20** (1): 151–77.

Buchanan, James M. and Gordon Tullock (1962) *The Calculus of Consent – Logical Foundations of Constitutional Democracy*, Ann Arbor; MI: University of Michigan Press.

Burrell, Gibson and Gareth Morgan (1979), *Sociological Paradigms and Organizational Analysis*, London: Heinemann.

Cohen, W. M. and D. A. Levinthal (1990), 'Absorptive Capacity: A New Perspective on Learning and Innovation', *Administrative Science Quarterly*, **35** (1): 128–52.

Donaldson, Lex (1995), *American Anti-Management Theories of Organization: A Critique of Paradigm Proliferation*, Cambridge: Cambridge University Press.

Grubel, H. G. and L. A. Boland (1986), 'On the Efficient Use of Mathematics in Economics: Some Theory, Facts and Results of an Opinion Survey', *KYKLOS*, **39** (3): 419–42.

Hayek, Friedrich A. von (1948), 'Economics and Knowledge', in Friedrich A. von Hayek (ed.), *Individualism and Economic Order*, Chicago, IL: University of Chicago Press, pp. 33–56.

Holmwood, J. and A. Stewart (1994), 'Synthesis and Fragmentation in Social Theory: A Progressive Solution', *Sociological Theory*, **12** (1): 83–100.

Kitcher, Philip (1993), *The Advancement of Science – Science without Legend, Objectivity without Illusions*, Oxford: Oxford University Press.

Knudsen, Christian (1993), 'Equilibrium, Perfect Rationality and the Problem of Self-Reference in Economics', in Uskali Mäki, Bo Gustafsson and Christian Knudsen (eds), *Rationality, Institutions and Economic Methodology*, London: Routledge, pp. 133–70.

Knudsen, Christian (2003), 'Pluralism, Scientific Progress and the Structure of Organization Studies', in Haridimos Tsoukas and Christian Knudsen (eds), *The Oxford Handbook of Organisation Studiess*, Oxford: Oxford University Press.

Koontz, H. (1961), 'The Management Theory Jungle', *Academy of Management Journal*, **4** (3): 174–88.

Koontz, H. (1980), 'The Management Theory Jungle Revisited', *Academy of Management Review*, **5** (2): 175–87.

Krugman, Paul (1996), 'How To Be a Crazy Economist', in Steven G. Medema and Warren J. Samuels (eds), *Foundations of Research in Economics: How Do Economists Do Economics?* Cheltenham: Edward Elgar, pp. 131–141.

Kuhn, Thomas S. (1970), *The Structure of Scientific Revolution*, Chicago, IL: University of Chicago Press.

Kuhn, Thomas S. (1977), *The Essential Tension*, Chicago, IL: University of Chicago Press.

Lakatos, Imre (1970), 'Falsification and the Methodology of Research Programmes', in Imre Lakatos and Alan Musgrave (eds), *Criticism and the Growth of Knowledge*, Cambridge: Cambridge University Press, pp. 91–196.

Laudan, Larry (1977), *Progress and Its Problem*, Berkeley, CA: University of California Press.

Mayr, E. (1976), *Evolution and the Diversity of Life*, Cambridge, MA: Harvard University Press.

Mintzberg, Henry, Bruce Ahlstrand and Joseph Lampel (1998), *Strategy Safari: A Guided Tour through the Wilds of Strategic Management*, New York, NY: The Free Press.

Mone, M. A. and W. McKinley (1993), 'The Uniqueness Value and Its Consequences for Organization Studies', *Journal of Management Inquiry,* **2** (3): 284–96.

Nelson, Richard R. and Sidney G. Winter (1982), *An Evolutionary Theory of Economic Change*, Cambridge, MA: The Belknap Press of Harvard University Press.

Pfeffer, Jeffrey (1982), *Organizations and Organization Theory,* Boston, MA: Pitman.

Pfeffer, J. (1993), 'Barriers to the Advance of Organizational Science: Paradigm Development as a Dependent Variable', *Academy of Management Review,* **18** (4): 599–620.

Poole, Marshall S. and Andrew H. Van de Ven (1989), 'Using Paradox to Build Management and Organization Theories', *Academy of Management Review,* **14** (4): 562–78.

Popper, Karl R. (1945), *The Open Society and Its Enemies,* London: Routledge & Kegan Paul.

Popper, Karl R. (1972), *Objective Knowledge,* Oxford: Clarendon Press.

Rawls, John (1971), *A Theory of Justice*, Cambridge MA: Harvard University Press.

Reed, Michael (1992), *The Sociology of Organization – Themes, Perspectives and Prospects,* New York, NY: Harvester Wheatsheaf.

Schumpeter, Joseph A. (1950), *Capitalism, Socialism and Democracy*, New York, NY: Harper.

Scott, W. Richard (1998), *Organizations: Rational, Natural and Open Systems,* Englewood Cliffs, NJ: Prentice-Hall.

Simon, Herbert A. (1962), 'The Architecture of Complexity', *Proceedings of the American Philosophical Society*, **196**: 467–82.

Van de Ven, Andrew H. (1999), 'The Buzzing, Blooming, Confusing World of Organization and Management Theory: A View from Lake Wobegon University', *Journal of Management Inquiry,* **8** (2): 118–25.

Van de Ven, Andrew H. and Marshall S. Poole (1988), 'Paradoxical Requirements for a Theory of Organizational Change', in Robert Quinn and Kim S. Cameron (eds) *Paradox and Transformation: Toward a Theory of Change in Organization and Management*, Cambridge, MA: Ballinger/Harper Row, pp. 19–64.

Whitley, R. (1983), 'The Structure and Context of Economics as a Scientific Field', *Research of the History of Economic Thought and Methodology*, **4**: 179–209.

Whitley, Richard (1984a), *The Intellectual and Social Organization of the Sciences*, Oxford: Clarendon Press.

Whitley, R. (1984b), 'The Development of Management Studies as a Fragmented Adhocracy', *Social Science Information*, **23**: 125–46.

Winter, Sidney G. (1975), 'Optimization and Evolution in the Theory of the Firm', in Richard H. Day and Melvin W. Reder (eds), *Rational Choice: The Contrast Between Economics and Psychology,* Chicago, IL: Chicago University Press, pp. 73–99.

Zammuto, R. F. and T. Connolly (1984), 'Coping with Disciplinary Fragmentation', *Organizational Behavior Teaching Review*, **9** (1): 30–7.

# 3. Residual categories and the evolution of economic knowledge

**Matthias Klaes**

## 1. INTRODUCTION[1]

Given an ongoing scientific research endeavour, one important question is which explanatory categories should be maintained, revised, or discarded. From the perspective of the individual scientist, this decision is approached in terms of 'what works' (cf. Knorr-Cetina 1981). Not least due to Popper ([1935] 1994; 1963), however, the fundamentally social nature of scientific knowledge has been acknowledged. Scientific knowledge is not so much the cumulative result of scientists acting individually as the result of their mutual interactions within a common discourse. If one follows Popper, the essence of this collective foundation of scientific knowledge is the institution of scientific criticism. Due to the fallibilist nature of knowledge, individual scientists can never be sure of their results. But collectively, the scientific community shares and maintains those theories which have, up to this point, survived all attempts at refutation.

How do these two perspectives, of the individual scientist and of the scientific community which sustains scientific knowledge, compare? Can we talk of scientific progress carried by the purposive actions of scientists, or do we need to analyse the collective properties of scientific knowledge in their own right? The present chapter approaches these questions by investigating theory change in the social sciences, and economics in particular. Instead of rehearsing well-known arguments regarding the respective relevance of individualism and collectivism in economics and social theory though, a slightly different angle will be pursued here. The focus of attention is on 'invisible hand' processes at work in the constitution of (social) scientific knowledge which lead to discursive formations of scientific thought over time, displaying properties beyond those intended by the practising scientists themselves. This shifts the overall question within the evolution of scientific knowledge to issues of stability and change.

As part of the Popperian legacy in the philosophy of science, questions of stability and change achieved prominence through the work of Imre Lakatos and his attempt to interpret the actual development of science in terms of problem shifts which could be appraised from a rational perspective. Like Thomas Kuhn (1970) and others before him, Lakatos drew attention to the fact that from a historical point of view, one can observe a surprising stability in the basic categories of scientific discourse. While Kuhn explained the stability with reference to the social elements which make up a scientific 'paradigm', Lakatos (1970) located it on the conceptual level by focusing on the heuristics which guide the development of a 'research programme'. These heuristics determine which paths are considered fruitful for further development – Lakatos's 'positive heuristics', and which are to be avoided – the 'negative heuristics'. As a result, one can distinguish between the 'hard core' of a research programme, and its attending 'protective belt'. In terms of the terminology introduced above, the hard core corresponds to those explanatory categories which are protected from change by means of negative heuristics. The protective belt is constituted by the remaining categories, which are subject to adjustment according to the positive heuristics.[2]

This provides a particular perspective on theory change. The distinction between hard core and protective belt directs our attention to the moves employed within a scientific discourse to react to phenomena in its domain of investigation which seem to call the hard core into question. Lakatos' study of the history of Euler's theorem on the number of vertices, edges and faces of a polyhedron, for example, provides us with a list of stratagems – 'monster-barring', 'exception-barring', and 'monster-adjustment' – which might be employed in such cases (cf. Lakatos 1976).[3] All these stratagems identify on the one hand explanatory concepts which are further pursued on the basis of the positive heuristics of the research programme, and on the other hand residual categories which delimit the scope of the theory.

Take a metaphor due to Dennis H. Robertson ([1923] [1928] 1943, p. 85) for example, which has gained currency in the theory of the firm. Robertson compares firms to 'islands of conscious power in this ocean of unconscious co-operation like lumps of butter coagulating in a pail of buttermilk'. Were one to develop this dualism one-sidedly, by concentrating on a liberalist market theory based on voluntary exchange, Robertson's islands of conscious power would become the residual. Tendencies of this kind can be found in the work of Ronald Coase (1937), whose theory of the firm rests on a differentiated picture of market exchange in which the costs of exchange play a pivotal role: Coase remains silent on the question of authority in intra-firm governance. 'Direction' occupies a central place in his distinction between firms and the market, but only as a residual category.

Residuals provide a very useful device for declaring parts of the domain of investigation of a discourse as lying beyond its explanatory ambitions. In this respect, they represent one instance of the principle of isolation which has been suggested as one of the prominent organising devices of economic theorising (cf. Mäki 1992; 1994; Schlicht 1977; 1985). In the terms of Lakatos' stratagems, isolation may be regarded as similar to 'exception-barring': it specifies the domain of application of a theory (cf. Lakatos 1976, p. 26).

The remainder of the chapter explores the dynamics associated with residual categories as instances of order in economic discourse which are unintended in two respects. First, these dynamics cannot be conceived as the intended result of the attempt by economists to manage their explanatory concepts. Second, they cannot be the intended results of a perspective on economics which gives economic knowledge a progressive and cumulative nature. The argument will be advanced on the basis of historical examples. In the following section, a fundamental residual category of economics will be explored. In the two succeeding sections, two patterns of development will be distinguished and analysed. On the one hand, residual categories can give rise to a fragmentation of scientific discourse. On the other hand, they can motivate a synthesis. Both fragmentation and synthesis will be argued to exhibit the same underlying pattern of development.

## 2. RESIDUAL CATEGORIES IN ECONOMICS

Let me illustrate the dynamic role of residual categories by turning to an example from the recent history of economics. Arguably, one of the most prominent isolating moves in economic methodology is due to Carl Menger's (1883) attack on the 'younger' German historical school, which opened the *Methodenstreit* (cf. Ritzel 1950). Menger argued for an 'exact' approach in economics, aiming at the formulation of exceptionless laws to be derived from the postulate of atomistic individuals who act self-interestedly in order to satisfy their wants in the best possible way. This reduction to the economic aspect of individual behaviour implies that statements based on these laws cannot generally be expected to correspond with the phenomena found in the domain of investigation of economics. Hence, Menger interprets economic laws derived from first principles as true representations of the *economic* aspects of the complexities of the real world.[4] Any discrepancies with the actual economy become thus attributable to the operation of extra-economic influences.

It would be useful to have a shorthand notion under which to subsume the thus isolated extra-economic influences. And indeed, late 19th century economics was in possession of such a device: the notion of friction. Frictions were commonly used by economists to address the exchange difficulties imposed by a barter economy, the lack of synchronisation between receipts and payments, and the impediments against the transfer of assets. A quote from Inglis Palgrave's *Dictionary of Political Economy* (see Davidson 1896, pp. 160–1) provides an interesting glimpse of the theoretical inventory of the discipline at the turn of the 19th century. Under the entry 'Friction in Economics', one finds the following:

> The disturbing effects of causes that are not economic, on the action of the causes that are strictly so called, may be regarded as an 'economic friction'. ... Not only the customs, but the vices, follies, and mistakes of men are accountable for economic friction. ... Economic friction may further be described as the opposition encountered by the movements of capital and the inability of labour to meet readily the demand for work; and generally by all the circumstances which prevent economic forces from bringing about their natural effects the instant they come into operation.

It is difficult to think of a more clear-cut instance of a residual category. The notion of friction provided a powerful residual category of the research programme of economics as an exact science, the positive categories of which address the economic aspects of the domain of investigation under study. This does not imply that the non-economic aspects are not important, only that they should be dealt with in another research programme. One candidate in particular comes to mind: the discipline of sociology. And indeed, there is at least one prominent 'sociologist' who defined the scope of economics and sociology on the basis of this dualism: Vilfredo Pareto.

In his *Trattato di Sociologia Generale* (cf. Pareto 1966), Pareto contrasts the category of 'logical action' – which he regards as one of the central categories of economic analysis – with the category of 'non-logical action'. Logical action is that action to which economic analysis applies. It refers to a framework which derives its conclusions from the premise that atomistic individuals act in an instrumentally rational way to fulfil their exogenously given wants. Pareto emphasises that this type of action refers only to a subset of the world of social phenomena.[5] A comprehensive analysis of social action needs to step beyond the logicality of action. In Pareto's system, this is achieved by introducing the category of non-logical action. Non-logical action is defined residually as those acts which cannot be subsumed under logical action.

Confining logical action to the realm of economic theory[6], Pareto concentrates in the *Trattato* on non-logical action. He studies non-logical action by focusing on the subjectively held 'theories' which inform the behaviour of actors in cases which fall outside the scope of instrumental rationality. From this latter perspective, non-logical action appears to be based on ignorance and error. In both cases, the actor fails to employ the means best suited, given the local circumstances, to achieve her ends. Her action appears to be irrational. Pareto divides the 'irrational' theories underlying non-logical action into two kinds: justifications of why an act is undertaken, and ideas on how it should be performed. He comes to the conclusion that in the majority of cases, actors make sense of their actions in terms of 'justice', 'God' or similar higher ends.

In general, these ends rest on a 'sentiment' that a particular state of affairs is *in itself* desirable, quite independent from its effect on the individual utility calculus.[7] In the realm of non-logical action, action is guided not by instrumental rationality but by customs and internalised norms of conduct. While logical action is based on the atomistic individual, non-logical action is an aggregate phenomenon which only becomes recognisable as the shared conduct of a group of individuals. One would be reluctant to speak of a norm if it was only followed by a single individual.[8]

Seen from Pareto's perspective, social theory is based on a fundamental dualism. Arguably, logical action constitutes a central explanatory category of economics, at least in its neoclassical guise, while non-logical action, as a residual category of economics, has largely remained in the territory of sociology (and institutional economists, for that matter). But if both types of action address relevant parts of a common domain of investigation, two interesting questions open up. First, how do these research programmes fare in their independent attempts to address this domain of investigation on the basis of just one of the categories of the dualism? Can one speak of a productive division of labour, or should one rather speak of self-defeating projects analogous to an attempt to decentralise processes of joint production? Second, what happens if practitioners who are dissatisfied with this intellectual decentralisation start turning aspects of the residual into positive explanatory categories?

Adopting a terminology due to Holmwood and Stewart's (1994) revealing study of the dynamics of dualism in social theory,[9] I will refer to the first response to cope with a dualism – intellectual division of labour – as the strategy of 'fragmentation'. The second – the attempt to endogenise the residual – will be called the strategy of 'synthesis'. Both synthesis and fragmentation are instances of Lakatossian moves to adjust the protective belt of a research programme. Fragmentation proceeds by treating one category of the dualism as an exogenous residual ('exception-barring'), while synthesis tries to engage in Parsons's carving-out of positive categories from this residual ('monster-adjustment'). The following two sections study instances of fragmentation and synthesis in the history of economics.

## 3. FRAGMENTATION

One particular branch of economics seems well suited to study how far the implications of the logical action research framework can be pushed. Ever since Adam Smith's ([1776] (1981): IV.ii.9) suggestion that the rational pursuit of economic self-interest, 'led by an invisible hand', is embedded in a harmonious order which furthers the public good, economists have tried to arrive at a comprehensive theoretical framework to express this harmony. From early on, this programme was also pursued in mathematical terms (cf. Isnard (1781), Cournot ([1838] 1938) and von Thünen ([1850] 1910)), and has – at least until recently – been regarded as the 'deepest scientific resource of economists' (Foley 1970: 276).

The canonical formulation of the issue from today's perspective is due to Léon Walras ([1874] 1926), who set out a system of equations describing the interrelation of the actions of self-interested rational individuals in a multi-market economy. In the 20th century, Walras' general equilibrium approach became regarded as unsatisfactory. Mathematically, specifying as many equations as there are unknowns is not sufficient for the system to have a solution. The system may in fact have no solution at all, or even infinitely many. Thus, the demonstration of the operation of Adam Smith's harmonious order became subsumed under the quest for a mathematical proof that Walras' system has an unequivocal solution. It is this research tradition of 'general equilibrium theory' in modern economics which most clearly exhibits the quixotic nature of trying to achieve explanatory progress by further developing the implications of the logical action premise.[10]

Incidentally, one of the first rigorous mathematical proofs of the existence of a general economic equilibrium appears to be due to a paper by Robert Remak (1929) entitled 'Kann die Volkswirtschaftslehre eine exakte Wissenschaft werden?', 'Can Economics Become an Exact Science?' Remak's paper was the result of his contacts with Ladislaus von Bortkiewicz, a Russian mathematical economist instrumental in establishing the German section of the Econometric Society. Among the young economists gathering around him in Berlin there was also Jacob Marschak, who would head the Cowles Commission in Chicago from 1943 to 1948. The Cowles Commission played a crucial role in breathing new life into the Walrasian research programme after World War II. However, the 'resurrection' of the Lausanne school took place in Vienna a generation earlier. In the early 1930s, Carl Menger, the son of the defender of economics as an exact science, set up a mathematical colloquium. The focus of the research activity was to replace Walras' system of equations with a mathematically more satisfactory approach, starting from Gustav Cassel's (1918) formulation of the Walrasian system.

An important advance was due to Frederik Zeuthen (1933), who suggested what today is known in linear programming as 'complementary slackness conditions' which allow us to reformulate inequalities in terms of strict equalities once one has introduced so-called 'slack' variables. Zeuthen was unable to show that the modified system had any solutions. Another participant in the colloquium, the young mathematician Abraham Wald (1936) took up Zeuthen's approach and was more successful, albeit under highly restrictive conditions. The most famous early existence proof is however due to John von Neumann (1937), who was teaching in Berlin from 1927 to 1929 and is likely to have had contacts with the Bortkiewicz group. With its ingenious application of the Brouwer fixed-point theorem, von Neumann's contribution foreshadowed the post-World War II Walrasian renaissance in Chicago and elsewhere, while being formulated not in the Vienna colloquium fashion but in the 'circular-flow' framework of Bortkiewicz and Remak which did not proceed from the neoclassical concept of scarcity but from the assumption that all commodities are reproducible within the system.

The rise of National Socialism put an end to the activities in Berlin and Vienna. Abraham Wald ended up at the University of Columbia, where he became one of the teachers of Kenneth J. Arrow's graduate years. Arrow (1995, p. 53) recalls that he learned in this way about the only partially solved 'existence' problem. But when he asked Wald about his work on the question, he got the reply that it was a very difficult problem: 'Coming from him, whose mathematical powers were certainly greater than mine, the statement was discouraging.' Help was however available from von Neumann's work. For the existence proof of his 1937 model, von Neumann had actually employed a generalised version of Brouwer's fixed-point theorem, which represented an extension of the minimax theorem for zero-sum two-person games he had developed earlier (von Neumann 1928). In the meanwhile, von Neumann had further developed his ideas on the application of the theory of convex sets to economic problems, through collaboration with Oskar Morgenstern, another colloquium member in the diaspora (von Neumann and Morgenstern 1944).

The English translation of this work served as an inspiration to the young mathematical economists like Arrow who turned to the existence problem after World War II (cf. Arrow [1951] 1983, p. 14). The von Neumann existence proof did not easily relate to the conventional categories of economic analysis. However, when John F. Nash Jr. (1950) applied a theorem equivalent to the one von Neumann had used to prove the existence of equilibria in game theory, Arrow, Gerard Debreu, and Lionel McKenzie all realised independently of each other how the existence problem in general equilibrium theory could be solved (cf. Arrow and Debreu 1954; McKenzie 1954).

As this chapter so far suggests, the historical texture surrounding the existence proofs is rich, and the material lends itself to the writing of 'heroic' historical accounts, in which a century-old struggle finally comes to rest, thanks to the successful application of formalist rigour. In fact, the problem – that of formal existence – was much less history-laden than is often suggested. While it is true to say that Walras was interested in the existence of an equilibrium, he was more concerned with an equilibrium in the actual economy than with the properties of a formal mathematical system (cf. van Daal and Jolink 1993, pp. 111–12). Mathematical existence came up as a topic in the Vienna colloquium, but that was in the 1930s. Given that Arrow ([1973] 1983, p. 217) describes the work of Tjalling Koopmans (1951) as 'an essential precondition for our [Debreu and Arrow's] studies', one might even regard the existence proofs as solutions of a set of problems closely related to the context of World War II.[11]

Did the answer to this fairly recent problem of formal existence advance in any way the Smithian case of a harmonious order springing from what in the terminology of this present chapter is called 'logical action'? Arrow and Debreu (1954: 60) argue that 'The view that the competitive model is a reasonably accurate description of reality ... presupposes that the equations describing the model are consistent with each other.' As mathematical economists were well aware, existence of equilibrium is just one aspect of this consistency. Two other properties, uniqueness and stability of equilibrium, are just as important. If an equilibrium is not stable, it cannot be reached except by 'jumping right into it', and any small perturbation will drive the economy away from it again. Furthermore, both uniqueness and stability are necessary for the applicability of one of the fundamental tools of analysis of modern economics, the exercise of comparative statics.

As subsequent work in general equilibrium theory has revealed, neither of these two properties can be assured by assumptions of the same order of generality as the ones employed for establishing existence. These limitations had already become apparent in Wald's (1936) early work, but were confirmed by the post-war formalists (Arrow, Block, and Hurwicz 1959; Scarf 1960; Sonnenschein 1972). In particular Sonnenschein's paper convinced the profession that the very general assumptions made at the level of individuals failed to place sufficient restrictions on the behaviour of aggregate excess demand to obtain the desired results (cf. Ingrao and Israel (1990, pp. 313–62) for details and qualifications.).

Kirman (1989) has located the source of this impasse in the premise of atomistic individuals acting independently of each other. He finds it hardly convincing to blame two other potential culprits, the assumption of optimising behaviour, or the 'straightjacket' of the mathematical framework within which the negative results are formulated. General equilibrium reasoning is based on individuals acting according to the principles of logical action. This does not preclude interaction between individuals. However, this interaction is restricted to the signals given by equilibrium market prices. Any other dependencies are regarded as 'externalities' which upset the working of the competitive system. Kirman points out that as soon as one allows for some interdependence of the demand behaviour of individuals, the set of aggregate excess demand functions becomes limited.

Hence, if pursued to the limit, the project of fragmentation followed in the economic approach gets drawn into its residual, the category of non-logical action. As has been mentioned above, non-logical action can only be understood as an aggregate phenomenon. What the research programme of general equilibrium theory points to is precisely the necessity of some structure at some level of aggregation beyond the single individual: 'If we are to progress further we may well be forced to theorise in terms of groups who have collectively coherent behaviour' (Kirman 1989: 138).

It is worth reflecting on the overall pattern of development of the general equilibrium discourse. Phrasing it in terms of the posing and solving of the existence problem seems to indicate a linear progression. As it were, the restriction to logical action has allowed us to formulate the metaphor of the invisible hand in more rigorous terms, which in turn brought forth a conclusive corroboration of its underlying conceptual apparatus. The impasse in the questions of uniqueness and stability of equilibrium, together with Kirman's observation, add a cyclical dimension to this story, as the strategy of fragmentation ends up calling forth a strategy of synthesis. Progress seems possible only if aspects of the residual are incorporated in the economic project of logical action.

## 4. SYNTHESIS

After having explored the consequences of pursuing a division of labour in the social sciences on the basis of the strategy of fragmentation, let us now explore the results of attempting a synthesis between explanatory categories in economics and their residual. Above, the notion of friction was cited as a prime example of a residual category in economics. Yet, in spite of the general acceptance of this category at the turn of the 20th century, some economists expressed their dissatisfaction with what they regarded as an ad hoc device designed to keep difficult but important issues outside the realm of economic analysis. In a classic article, Hicks (1935) attacked the use of the notion of friction in early neoclassical monetary theory. The transactions and cash-balance approaches of Irving Fisher and the Cambridge school treated the amount of cash people held for transacting purposes as an exogenous institutional factor. Hicks (1935, p. 6) called the adherents of these schools the 'great evaders'. He demanded that the marginalist paradigm be extended to monetary theory.

What was needed was an analysis of the choice of an individual economic agent between holding money or other assets. To make this step, the notion of friction had to be converted into one of the positive categories of neoclassical economics. Hicks suggested addressing the frictions of monetary theory in terms of costs. People hold cash because of the transaction charges and the time which would be involved if one withdrew money for every payment separately. Largely in response to his initiative, the 1950s saw the gradual emergence of an ever-growing discourse centred around a concept which has become well-known not just in economics but also in neighbouring social sciences: the concept of transaction costs (cf. Klaes 2000a).

In monetary economics, the early proponents were Marschak's (1950) general equilibrium model of an economy with transaction costs, and Baumol (1952) and Tobin's (1956) so called 'cash-balance' models of the transactions demand for cash. While Marschak's article was largely neglected, the work of Baumol and Tobin gave rise to a large literature, forming part of the post-war movement of neo-Keynesianism which tried to reconcile Keynes' General Theory with neoclassical theory. In this initial period, transaction costs were predominantly understood as brokerage charges or the costs of investing in financial markets in general. But besides their role as a positive category to make some aspects previously subsumed under the notion of friction amenable to economic analysis, there was a second tendency in employing transaction costs: as the new label for the residual. This tendency can be traced back to the so-called Modigliani-Miller theorems on corporate finance, which were formulated to apply to a world free of 'transaction costs' (Modigliani and Miller 1958; 1959; 1963; Miller and Modigliani 1961).

By the mid-1960s, the concept assumed increasingly broader interpretations in the process of being taken up in other sub-disciplines. The conceptual migration of transaction costs from their original context into other domains started with contributions by Marschak (1950), Hymer ([1960]

1976), and Malmgren (1961) on the economics of information and the theory of the firm. This migration can be nicely observed in the first debate on the substance of the concept of transaction costs itself, which took place between Kenneth Arrow and the two health economists Dennis Lees and Robert Rice (cf. Klaes 2000b). Prompted by an attack by Lees and Rice (1965) on his analysis of health insurance, Arrow (1963; 1965) defended himself by broadening the concept beyond the narrow sense it had acquired in the monetary and financial literature, and which he had initially endorsed himself (Arrow, Cagan and Friend 1963). This eventually led him to embrace it as a shorthand expression of market imperfections in general, and ultimately as a category reaching beyond the market towards other forms of resource allocation (Arrow [1969] 1983).

Around the same time, authors like H. Miller (1965) suggested interpreting transaction costs as the cost of time, while Alchian and Allen's (1964) influential textbook *University Economics* introduced an interpretation of transaction costs which emphasised the aspect of contracting, and especially the enforcement of property rights. The textbook turned into a prominent tool for introducing lawyers to the emerging literature on the economics of property rights. This cross-fertilisation was also propelled by lawyers like Calabresi (1968) and Liebeler (1968) who worked in the new field of law and economics. As a by-product of these developments, transaction costs diffused into the externalities literature (McKean 1970; Mishan 1971), and the related field of environmental economics (Crocker 1971; 1973; Daly and Giertz 1975).

In the 1970s, probably the most important impetus to the further diffusion of the concept was due to the formation of a transaction cost 'school' around Armen Alchian at UCLA (cf. Cheung 1969; Demsetz 1968), the work of Oliver Williamson (1970; 1975) on corporate governance, and the emergence of a 'new' institutional economics.[12] The UCLA school exerted a strong influence on transaction cost arguments in economic history (cf. Davis and North 1971).

Thus, by the mid-1970s, the concept of transaction costs was used in a whole variety of different fields in economics, finance, business studies, law, and economic history. At the same time, however, concern grew about its loose nature and it attracted criticism for being too 'vague', constituting a 'catch-all' term, not having been 'clearly articulated' yet, or not having been properly understood (Cheung 1974; Krier and Montgomery 1973; Miller 1965; Niehans 1969; Shapiro 1971). It is interesting to observe that to a large extent, this criticism echoes the original attack of Hicks (1935) on the use of the notion of friction in economics. Furthermore, the critical voices are not confined to the 1970s but have continued to plague transaction cost discourse to the present (cf. Allen 1991; Davies 1986; Dixit 1996).

While one should be careful not to over-generalise, there is a striking tendency in modern economics to use the term 'transaction costs' synonymously with the 19th century notion of friction. It has assumed the place of a general metaphor to delineate the applicability of orthodox economics: 'Transaction cost analysis ... is appropriate for studying the frictions in the system which may prevent the implications of received microtheory from going through' (Williamson, 1974: 1495).[13]

Taking a step back from this historical case study, two observations can be made. On the one hand, it seems that residual categories can furnish material for the conceptual development of a research programme. This is illustrated by Hicks's suggestion translating frictions into the economic concept of cost. A potential attack – 'look, economics cannot help us a bit in understanding all these important phenomena which it subsumes under the notion of friction' – is accommodated by partially endogenising the residual. What seemed a shortcoming of the logical action framework has become converted into yet another strength. Calling frictions a cost further supports the categories in the core of the programme (a clear instance of Lakatos's stratagem of 'monster-adjustment').

This mechanism of theory development would fit well into a teleological account of scientific progress: theories change as a result of a 'carving out' of positively defined categories from the set of residual categories (Parsons [1937] [1949] 1968, p. 18). Attempts to endogenise the residual would thus represent special instances of the general problem-solving nature of science. For Parsons, the process of carving out is an expression of the asymptotic goal of science to eliminate residual categories altogether. If this cannot be achieved within a single theoretical framework, one should at least hope that what remains as the residual category in one framework becomes elucidated by the positive category of another framework.

The story of the carving out of transaction costs from the notion of friction does not fully support such a linear account of theoretical progress, however. In fact, it might encourage a much more pessimistic conclusion. The fate of transaction costs in the literature of modern economics reads like the history of the quixotic struggle of the discipline to endogenise one of the most pervasive residual categories of the neoclassical heritage (Klaes 2000a). While a shift from friction to transaction costs seems progressive, the subsequent dissolution of transaction costs into a 'higher' form of friction would rather suggest a circular pattern.

One could argue that the circular pattern, while present on the linguistic level, should not be taken as indicative of the theoretical development sustained by the discourse studied. The dissolution of transaction costs into vagueness might be regarded as a linguistic 'smoke-screen' which should not distract us from possible theoretical advances. One could for example argue that Williamson has furthered our theoretical understanding of institutional governance, and that this development has proceeded quite independently of his loose use of the concept of transaction costs. In fact, much of the explanatory import of Williamson's framework rests on the concept of asset specificity, and not on the concept of transaction costs.

Yet, the circular pattern can also be observed at the level of theories, not in transaction-cost economics however but in the general equilibrium literature. For example, the attempts of the 1970s to incorporate the concept of transaction costs into general equilibrium models opened up fundamental questions regarding the conventional interpretation of efficiency (cf. Klaes 2000b). Similarly, attempts to measure transaction costs quantitatively reveal that the concept 'uproots' its own foundations in the neoclassical cost concept: significant levels of transaction costs in the economy represent at the same time significant deviations from the doctrine that costs should be interpreted as opportunity costs (cf. Klaes 1996).

## 5. CONCLUSION

Imre Lakatos suggested that the remarkable stability which can be observed in scientific research programmes is due to persistent efforts of scientists to maintain a given hard core even in times when recurring difficulties become apparent. Scientists achieve this stability by clever manipulations of the protective belt of their theoretical frameworks. One lesson which can be learned from the historical case studies presented above is that these manipulations may lead to particular discursive dynamics which stand in contrast to a linear evolution of scientific knowledge, both in terms of the problem-solving activity of individual scientists, and in terms of the progressive and cumulative nature of scientific knowledge. In the discourses studied, social scientists have been aware of some difficulties associated with the dualisms at the heart of their theories. In response, they have followed either a strategy of fragmentation, trying to develop the component categories of the dualism separately by isolating the residual; alternatively, they have adopted a strategy of synthesis, in their attempts to carve out positive categories from the residual.

In both cases, the discourses exhibited an astonishing resilience to attempts to fruitfully develop the implications arising from the accepted dualism. In general equilibrium theory, progress in the project of fragmentation was marred by the impasses encountered in the questions of stability and uniqueness, which pointed to the significance of the residual. Synthetic attempts to incorporate aspects of the residual into the positive categories of the theoretical framework by translating frictions into costs were equally frustrated. The resulting pattern of development was tentatively described as 'cyclical' above. As a process not consciously set in motion by the individual scientist, it lends an unintended stability to the hard core of a research programme, compared to the intended stability by explicit adjustments in the protective belt.

Further research is needed to establish whether the identified patterns are indeed best described with reference to a cycle. The cycle metaphor would require some repetition, for example, with the caveat that historical processes never quite repeat themselves on the token level. The various impasses described above would need to be followed by new episodes of optimistic endeavours to achieve progress, following different lines of development while still maintaining the original dualism. One could think of some plausible ways to argue that the impasse in general equilibrium theory has repeated itself on another level in the rise and stagnation of (non-evolutionary) game-theory in the economics of the past two or three decades. Similarly, there are some indications that the discourse of social capital, which represents a carving-out attempt similar to the one of transaction costs, is running into difficulties, too.

Nevertheless, the cycle metaphor, possibly broadened into W. B. Yeats ([1925] 1978) the imagery of 'gyres' lends itself well to illustrating the systemic effects present in the dynamics of the development of scientific discourse. Evidence of circularity, albeit only of a historical nature, calls into question linear perspectives of scientific progress. While the problem-solving attempts of the individual researcher remain a crucial factor in the processes which sustain scientific knowledge, there is no guarantee that the aggregation of these attempts over time results in an outcome the scientist, or traditional philosopher of science, would wish for. It is worth stressing that this result does not amount to a criticism of scientific practice. But the possibility of spontaneously emerging patterns of scientific discourse associated with particular manipulations of explanatory categories provides a caveat to any research tradition in danger of becoming too proud of its stability.

## NOTES

1 Thanks to the organisers and the participants of the workshop 'Science as Spontaneous Order' (Klitgaarden, 15–19 January 2000) for their encouragement and constructive criticism. I would also like to thank Emrah Aydinonat, Kurt Dopfer, Maarten Janssen, Albert Jolink, Arjo Klamer, David Knights, Nick Lee, Marco Lehmann-Waffenschmidt, Uskali Mäki, Rolland Munro, Sitki Yurekli and participants of the 2000 Wartensee Workshop on Evolutionary Economics, University of St. Gallen, the CSTT seminar at Keele University, and the EIPE 1999 work-in-progress seminar at Erasmus University for their critical comments and helpful suggestions on earlier stages of this work. I owe a particular debt to John Holmwood.

2 It is debatable whether Lakatossian hard cores do in fact offer a useful way of identifying research programmes. Historical research is plagued by the problem of unambiguously defining hard cores, for example. What is required for the subsequent line of reasoning, however, is not a consistent delineation of hard cores but the relative stability of some explanatory categories which are, in a loose sense, characteristic of a given line of research.
3 Briefly, 'monster-barring' refers to an ad hoc redefinition of categories to accommodate cases which conflict with the theory. Exception-barring refers to revising the domain of validity of the theory systematically. On monster-adjustment, let us cite directly from Lakatos' Socratic dialogues: 'Monsters don't exist, only monstrous interpretations. ... My method is therapeutic: where you – erroneously – "see" a counterexample, I teach you how to recognise – correctly – an example.' (Lakatos 1976, p. 31).
4 This interpretation of Menger has recently been strengthened by Mäki (1997) who argues that Menger employs a realist notion of universals.
5 Note the different form of isolation involved, which can be made visible with Uskali Mäki's (1994) distinction between horizontal and vertical isolation. Economics as an exact science isolates the economic aspects vertically, while Pareto isolates logical action horizontally.
6 See Pareto's ([1896-97] 1964) *Cours d'Economie Politique*, which leaned on the general equilibrium framework of Léon Walras ([1874] 1926), his predecessor in the Chair of Economics at the University of Lausanne.
7 Pareto discusses these sentiments as a particular class of his category of 'residues'. This needs to be understood on the basis of his 'positivistic' (Parsons [1937] [1949] 1968, pp. 58, 187) theory of science. Scientific theories, which provide the standard on which to judge the logicality of action, are a combination of statements of fact and logical reasoning. Similarly, the theories underlying non-logical action can be broken down into manifestations of sentiments, and deductions drawn from these sentiments on the basis of sophisms and similar ways of reasoning. Pareto calls the first component 'residues' and the second 'derivations' (Pareto 1966, pp. 216–8; cf. Parsons [1937] [1949] 1968, p. 198).
8 This leaves open whether the shared nature of this conduct has bio-psychological or social origins. For the former position, see Schlicht (1998), on the latter, Bloor (1997). I will not go into the question here to what extent logical action may be able to account for normativity. Suffice it to note that examples such as the persistent custom of tipping in non-repeated anonymous encounters are not easy to accommodate in a framework based on logical action.
9 See also Holmwood and Stewart (1991). This is not the place to appraise Holmwood and Stewart's account and criticism of social theory (cf. Colomy 1993).
10 As the details of this episode of doctrinal history are well known (cf. Ingrao and Israel 1990) the remainder of this section follows closely the historical outline presented in Screpanti and Zamagni (1993, pp. 258–62; 340–55).
11 Koopmans developed his activity analysis – a form of what today is known as linear programming during World War II, while studying the logistics of ocean shipping (cf. Jolink 1999). Jolink points out that Koopmans recognised the relevance of his analysis for the question whether a central planner could efficiently allocate resources, which was one of the central topics of the debate on socialism of the 1930s. Seen from this angle, the subsequent existence proofs were not prompted by a study of markets but by a study of central planning. It is no coincidence that both Arrow ([1951] 1983) and Debreu (1951) first provided a new proof of the efficiency of competitive equilibrium, exploiting a reformulation of this problem in the terms of linear programming, before turning to the existence problem. For the impact of World War II on post-war economics in this context, cf. Mirowski (2002).
12 'New' because its adherents wanted to distance themselves from the American institutionalism of the inter-war period, which was in important respects highly critical of economic orthodoxy. While this 'old' institutionalism may be seen as trying to develop an alternative research programme, new institutional economics accepted the basic tenets of orthodoxy while working on the relaxation of some parts of its protective belt.
13 This point is developed in greater detail in Klaes (2000a). The increasing degree of polymorphicity (cf. Ryle 1951) in transaction cost discourse should however not be taken as a vice. Concepts may serve as umbrella notions which help stabilise a dissenting discourse

and the social formations which sustain it, thus forming an important ingredient in the emergence of rival research programmes.

## REFERENCES

Alchian, Armen A. and William R. Allen (1964), *University Economics*, Second printing, Belmont, CA: Wadsworth.

Allen, D. W. (1991), 'What Are Transaction Costs?', *Research in Law and Economics,* **14**: 1–18.

Arrow, Kenneth J. ([1951] 1983), 'An Extension of the Basic Theorems of Classical Welfare Economics', in Kenneth J. Arrow (ed.), *Collected Papers of Kenneth J. Arrow*, Vol. 2, Oxford: Blackwell, pp. 13–45.

Arrow, Kenneth J. (1963), 'Uncertainty and the Welfare Economics of Medical Care', *American Economic Review,* **53** (5): 941–73.

Arrow, Kenneth J. (1965), 'Uncertainty and the Welfare Economics of Medical Care: Reply (the Implications of Transaction Costs and Adjustment Lags)', *American Economic Review,* **55** (1/2): 154–8.

Arrow, Kenneth J. ([1969] 1983), 'The Organization of Economic Activity: Issues Pertinent to the Choice of Market versus Nonmarket Allocation', in Kenneth. J. Arrow (ed.), *Collected Papers of Kenneth J. Arrow*, Vol. 2, Oxford: Blackwell, pp. 133–55.

Arrow, Kenneth J. ([1973] 1983), 'General Economic Equilibrium: Purpose, Analytic Techniques, Collective Choice', in Kenneth J. Arrow (ed.), *Collected Papers of Kenneth J. Arrow*, Vol. 2, Oxford: Blackwell, pp. 199–227.

Arrow, Kenneth J. (1983), 'Introduction to "An Extension of the Basic Theorems of Classical Welfare Economics"', in Kenneth J. Arrow (ed.), *Collected Papers of Kenneth J. Arrow*, Vol. 2, Oxford: Blackwell, pp. 13–4.

Arrow, Kenneth J. (1995), 'Kenneth J. Arrow', in William Breit and Roger W. Spencer (eds), *Lives of the Laureates: Thirteen Nobel Economists*, Cambridge, MA: MIT Press, pp. 43–58.

Arrow, Kenneth J., H. D. Block and L. Hurwicz (1959), 'On the Stability of the Competitive Equilibrium, II', *Econometrica,* **27** (1): 82–109.

Arrow, Kenneth J., P. Cagan and I. Friend (1963), 'Comment on Duesenberry's "The Portfolio Approach to the Demand for Money and Other Assets"', *Review of Economics and Statistics,* **45** (1): 24–31.

Arrow, Kenneth J. and G. Debreu (1954), 'Existence of an Equilibrium for a Competitive Economy', *Econometrica,* **22** (3): 265–90.

Baumol, W. J. (1952), 'The Transactions Demand for Cash: An Inventory Theoretic Approach', *Quarterly Journal of Economics,* **66** (4): 545–56.

Bloor, David (1997), *Wittgenstein: Rules and Institutions*, London: Routledge.

Calabresi, G. (1968), 'Transaction Costs, Resource Allocation and Liability Rules – A Comment', *Journal of Law and Economics,* **11** (1): 67–73.

Cassel, Gustav (1918), *Theoretische Sozialökonomie*, Leipzig: Winter.

Cheung, S. N. S. (1969), 'Transaction Costs, Risk Aversion, and the Choice of Contractual Arrangements', *Journal of Law and Economics,* **12** (1): 23–42.

Cheung, S. N. S. (1974), 'A Theory of Price Control', *Journal of Law and Economics,* **17** (1): 53–71.

Coase, R. H. (1937), 'The Nature of the Firm', *Economica,* **4** (16): 386–405.

Colomy, P. (1993), 'Explanation and Social Theory, by John Holmwood and Alexander Stewart', *Contemporary Sociology,* **22** (2): 288–9.

Cournot, Augustin ([1838] 1938), *Recherches sur les Principes Mathématiques de la Théorie des Richesses*, Paris: Rivière.

Crocker, T. D. (1971), 'Externalities, Property Rights, and Transaction Costs', *Journal of Law and Economics,* **14**: 451–64.

Crocker, T. D. (1973), 'Contractual Choice', *Natural Resources Journal,* **13**: 561–77.

Daly, G. and J. F. Giertz (1975), 'Externalities, Extortion, and Efficiency', *American Economic Review,* **65** (5): 997–1001.

Davidson, M. G. (1896), 'Friction in Economics', in R. H. Inglis Palgrave (ed.), *Dictionary of Political Economy*, Vol. 2, London: Macmillan, pp. 160–1.

Davis, Lance E. (1986), 'Comment', in Stanley L. Engerman and Robert E. Gallman (eds), *Long-Term Factors in American Economic Growth.* Chicago, IL and London: Universty of Chicago Press, pp. 149–61.

Davis, Lance E., and Douglass C. North (1971), *Institutional Change and American Economic Growth*, Cambridge: Cambridge University Press.

Debreu, G. (1951), 'The Coefficient of Resource Utilization', *Econometrica* **19** (3): 273–92.

Demsetz, H. (1968), 'The Cost of Transacting', *Quarterly Journal of Economics,* **82** (1): 33–53.

Dixit, Avinash K. (1996), *The Making of Economic Policy: A Transaction-Cost Politics Perspective*, Cambridge, MA: MIT Press.

Foley, D. K. (1970) 'Economic Equilibrium with Costly Marketing', *Journal of Economic Theory,* **2** (3): 276–91.

Hicks, J. R. (1935), 'A Suggestion for Simplifying the Theory of Money', *Economica,* **2** (5): 1–19.

Holmwood, John and Alexander Stewart (1991), *Explanation and Social Theory*, London: Macmillan.

Holmwood, J. and A. Stewart (1994), 'Synthesis and Fragmentation in Social Theory: A Progressive Solution', *Sociological Theory,* **12** (1): 83–100.

Hymer, Stephen H. ([1960] 1976), *The International Operations of National Firms: A Study of Foreign Direct Investment*, Cambridge, MA: MIT Press.

Ingrao, Bruna and Giorgio Israel (1990), *The Invisible Hand: Economic Equilibrium in the History of Science*, Cambridge, MA: MIT Press.

Isnard, Achylle-Nicolas (1781), *Traité des Richesses*, London and Lausanne: Grasset.

Jolink, Albert (1999), 'The Travelling Salesman Returns from the War: Tjalling Koopmans and War-Time Studies for Peace-Time Applications', *EIPE Workshop 'Economists at War'*, Erasmus Institute for Philosophy and Economics, Erasmus University, Rotterdam, 21 April.

Kirman, A. (1989), 'The Intrinsic Limits of Modern Economic Theory: The Emperor Has No Clothes', *Economic Journal,* **99** (395): 126–39.

Klaes, M. (1996), 'Transaction Costs as Opportunity Costs: A Conceptual Analysis', *Annual Conference of the European Association for Evolutionary Political Economy (EAEPE)*, Antwerp, 7–9 November.

Klaes, M. (2000a), 'The History of the Concept of Transaction Costs: Neglected Aspects', *Journal of the History of Economic Thought,* **22** (2): 191–216.

Klaes, M. (2000b), 'The Birth of the Concept of Transaction Costs: Issues and Controversies', *Industrial and Corporate Change,* **9** (4): 567–93.

Knorr-Cetina, Karin D. (1981), *The Manufacture of Knowledge – An Essay on the Constructivist and Contextual Nature of Science*, Oxford and New York, NY: Pergamon.

Koopmans, Tjalling C. (1951), 'Analysis of Production as an Efficient Combination of Activities', in Tjalling C. Koopmans (ed.), *Activity Analysis of Production and Allocation*, New York, NY: Wiley, pp. 33–97.

Krier, J. E. and D. Montgomery (1973), 'Resource Allocation, Information Cost and the Form of Government Intervention', *Natural Resources Journal,* **13**: 89–105.

Kuhn, Thomas S. (1970), *The Structure of Scientific Revolutions*, Second edition. Chicago, IL: University of Chicago Press.

Lakatos, Imre (1970), 'Falsification and the Methodology of Research Programmes', in Imre Lakatos and Alan Musgrave (eds), *Criticism and the Growth of Knowledge,* Cambridge: Cambridge University Press, pp. 91–196.

Lakatos, Imre (1976), *Proofs and Refutations: The Logic of Mathematical Discovery,* Cambridge: Cambridge University Press.

Lees, D. S. and R. G. Rice (1965), 'Uncertainty and the Welfare Economics of Medical Care: Comment', *American Economic Review,* **55** (1/2): 140–54.

Liebeler, W. J. (1968), 'Toward a Consumers Antitrust Law: The Federal Trade Commission and Vertical Mergers in the Cement Industry', *UCLA Law Review,* **15**: 1153–202.

McKean, R. N. (1970), 'Product Liability: Implications of Some Changing Property Rights', *Quarterly Journal of Economics,* **84** (4): 611–26.

McKenzie, L. W. (1954), 'On Equilibrium in Graham's Model of World Trade and Other Competitive Systems', *Econometrica,* **22** (2): 147–61.

Mäki, Uskali (1992), 'On the Method of Isolation in Economics', in Craig Dilworth (ed.), *Intelligibility in Science*, Amsterdam: Rodopi, pp. 317–51.

Mäki, Uskali (1994), 'Isolation, Idealization and the Truth in Economics', in Bert Hamminga and Neil De Marchi (eds) *Idealization VI: Idealization in Economics,* Amsterdam: Rodopi, pp. 147–68.

Mäki, Uskali (1997), 'Universals and the *Methodenstreit*: A Re-examination of Carl Menger's Conception of Economics as an Exact Science', *Studies in the History and Philosophy of Science,* **28**: 475–95.

Malmgren, H. B. (1961), 'Information, Expectations and the Theory of the Firm', *Quarterly Journal of Economics,* **75** (3): 399–421.

Marschak, J. (1950), 'The Rationale of the Demand for Money and of "Money Illusion"', *Metroeconomica,* **2**: 71–100.

Menger, Carl (1883), *Untersuchungen über die Methode der Socialwissenschaften, und der politischen Oekonomie insbesondere*, Leipzig: Duncker and Humblot.

Miller, H. L. (1965), 'Liquidity and Transaction Costs', *Southern Economic Journal,* **32** (1): 43–8.

Miller, M. H. and F. Modigliani (1961), 'Dividend Policy, Growth, and the Valuation of Shares', *Journal of Business,* **34** (4): 411–33.

Mirowski, Philip (2002) *Machine Dreams: Economics becomes a Cyborg Science,* New York, NY: Cambridge University Press.

Mishan, E. J. (1971), 'Pangloss on Pollution', *Swedish Journal of Economics,* **73**: 113–20.

Modigliani, F. and M. H. Miller (1958), 'The Cost of Capital, Corporate Finance, and the Theory of Investment', *American Economic Review,* **48** (3): 261–97.

Modigliani, F. and M. H. Miller (1959), 'The Cost of Capital, Corporation Finance, and the Theory of Investment: Reply', *American Economic Review,* **49** (4): 655–69.
Modigliani, F. and M. H. Miller (1963), 'Corporate Income Taxes and the Cost of Capital: A Correction', *American Economic Review,* **53** (3): 433–43.
Nash, J. F. (1950), 'The Bargaining Problem', *Econometrica,* **18** (2): 155–62.
Niehans, J. (1969), 'Money in a Static Theory of Optimal Payments Arrangements', *Journal of Money, Credit, and Banking,* **1** (4): 706–26.
Pareto, Vilfredo (1966), *Sociological Writings,* S. E. Finer (ed.), D. Mirfin (transl.), London: Pall Mall.
Pareto, Vilfredo ([1896-97] 1964), *Cours d'Économie Politique*, Georges H. Bousquet and Giovanni Busino (eds), Geneva: Droz.
Parsons, Talcott ([1937] [1949] 1968), *The Structure of Social Action*, 2 vols., New York, NY and London: The Free Press and Collier-Macmillan.
Popper, Karl R. ([1935] 1994), *Logik der Forschung*, Tenth edition, Tübingen: Mohr.
Popper, Karl R. (1963), *Conjectures and Refutations*, London: Routledge.
Remak, R. (1929), 'Kann die Volkswirtschaftslehre eine exakte Wissenschaft Werden?', *Jahrbücher für Nationalökonomie und Statistik,* **131** (Third series no. 76): 703–35.
Ritzel, Gerhard (1950), *Schmoller Versus Menger*, Frankfurt-am-Main: Enz and Rudolph.
Robertson, Dennis H. ([1923] [1928] 1943), *The Control of Industry*, London and Cambridge: Nisbet and Cambridge University Press.
Ryle, Gilbert (1951), Thinking and Language, *Collected Papers*, Vol. 2, London: Hutchinson, pp. 258–71.
Scarf, H. (1960), 'Some Examples of Global Instability of the Competitive Equilibrium', *International Economic Review,* **1** (3):157–72.
Schlicht, Ekkehart (1977), *Grundlagen der öknomischen Analyse*, Reinbek: Rowohlt.
Schlicht, Ekkehart (1985), *Isolation and Aggregation in Economics*, Berlin: Springer.
Schlicht, Ekkehart (1998), *On Custom in the Economy*, Oxford: Clarendon.
Screpanti, Ernesto and Stefano Zamagni (1993), *An Outline of the History of Economic Thought*, Oxford: Clarendon Press.
Shapiro, H. T. (1971), 'The Efficacy of Monetary and Fiscal Policies [Comment]', *Journal of Money, Credit, and Banking,* **3** (2): 550–4.
Smith, Adam ([1776] 1981), *An Inquiry Into the Nature and Causes of the Wealth of Nations*, in R. H. Campbell, A. S. Skinner, W. B. Todd (eds) Vol. 1, Indianapolis, IN: Liberty Classics.
Sonnenschein, H. (1972), 'Market Excess Demand Functions', *Econometrica,* **40** (3): 549–63.
Tobin, J. (1956), 'The Interest-Elasticity of Transactions Demand for Cash', *Review of Economics and Statistics,* **38** (3): 241–7.
van Daal, Jan and Albert Jolink (1993), *The Equilibrium Economics of Léon Walras*, London: Routledge.
von Neumann, John (1928), 'Zur Theorie der Gesellschaftsspiele', *Mathematische Annalen,* **100**: 295–320.
von Neumann, John (1937), 'A Model of General Economic Equilibrium', *Review of Economic Studies*, transl., 1945–46, **13** (33): 1–9.
von Neumann, John and Oskar Morgenstern (1944), *Theory of Games and Economic Behavior*, Princeton, NJ: Princeton University Press.

von Thünen, Johann-Heinrich ([1850] 1910), *Der isolierte Staat*, Second part, H. Waentig (ed.), Jena: Fischer.

Wald, A. (1936), 'Über einige Gleichungssyteme der mathematischen Ökonomie', *Zeitschrift für Nationalökonomie*, **7**: 637–70.

Walras, M. E. Léon ([1874] 1926), *Élements d'Économie Politique Pure*, revised and augmented edition, Paris: Pichon and Durand-Auzias.

Williamson, Oliver E. (1970), *Corporate Control and Business Behavior*, Englewood Cliffs, NJ: Prentice-Hall.

Williamson, Oliver E. (1974), 'The Economics of Antitrust: Transaction Cost Considerations', *University of Pennsylvania Law Review*, **122**: 1439–96.

Williamson, Oliver E. (1975), *Markets and Hierarchies: Analysis and Anti-Trust Implications: A Study in the Economics of Internal Organization*, New York, NY: The Free Press.

Yeats, W. B. ([1925] 1978), *A Vision*, G. M. Harper and W. K. Hood (eds), London: Macmillan.

Zeuthen, F. (1933), 'Das prinzip der Knappheit, technische Kombination, und ökonomische Qualität', *Zeitschrift für Nationalökonomie*, **4**: 1–24.

# 4. Evolutionary, constructivist and reflexive models of science

**Finn Collin**

## 1. INTRODUCTION

In this chapter, I compare the evolutionary model of the growth of science with the currently popular social constructivist models. I try to show that the evolutionary model has a number of advantages over the latter models. In the first place, the evolutionary model respects the truism that a theory (of natural science) survives in part because it fits objective reality, just as a living organism survives in part because it is adapted to its physical surroundings. This truism tends to be denied by social constructivists, or at least to be accorded a negligible significance.

Secondly, the evolutionary model, as I shall understand it here in a minimalist sense, aspires to explain only the fortunes of the competing entities once they have arisen, but not their genesis. Thus, an evolutionary theory of science will not try to explain the process through which hypotheses come to be formulated, but only their reception once they have appeared. In this, I shall argue, it is more judicious than certain varieties of social constructivism, which try to explain the very genesis of hypotheses. This ambition faces insurmountable problems since, if 'explanation' is taken in the rigorous sense intended by at least certain social constructivists, the generation of such explanations would require some sort of sociological super-theory which would encompass every theory of natural science.

It is necessary, however, to augment the evolutionary model so as to include, among the parameters against the background of which theories compete, also the users of the theories, the scientific community. This feature has no direct counterpart in evolutionary theories in biology.[1] The need for this expansion makes it possible for social constructivists to boast that their theories are more powerful than evolutionary theories, since they address an issue on which evolutionary theories are silent, namely concerning the nature of the factors determining the reception of theories in the scientific community. To account for this, such societal categories as interests or ideology are invoked. I try to show that this introduces an element of relativism in constructivist theories, which proves troublesome. The application of such theories will tend to change the conditions of application in ways that cannot be encompassed within those very theories.

Thus, the final conclusion will be that the critical comparison of evolutionary and social constructivist conceptions of science fails to produce a clear winner, but eventually points beyond both kinds of theory towards a superior conception which I term *reflexive* theories.

## 2. THE EVOLUTIONARY MODEL OF SCIENCE

According to an evolutionary model of science, the fate of a scientific theory is decided by its degree of 'adaptation' to the reality it is designed to capture. The governing principle is 'survival of the fittest', and the 'fittest' theory is literally the one that fits reality the best. This notion of 'fitting' or 'adaptation' need not be taken in the strong sense of 'representing' or even 'mirroring' the structure of reality; the notion of adaptation is consistent with a radical empiricist notion of science, such as Bas van Fraassen's 'constructive empiricism' (van Fraassen 1980), where a theory is only required to be in accordance with observational recordings. Indeed, the evolutionary model is even consistent with a blatantly instrumentalist interpretation of scientific theories.

The evolutionary model of science shows some affinity with Popper's methodology of 'eliminative induction', and was indeed endorsed by Popper in some of his later works (Popper 1972; 1974). For further sources of the evolutionary model of science, see for instance Radnitzky and Bartley (1987). A biological species wins a crucial victory in the struggle for survival when its competitors die out, in the same way that a scientific theory establishes itself when its rivals are falsified and abandoned. A scientific theory is not positively proven true when it does better than a rival on some particular test; it may be falsified by the very next one. Similarly, the fact that a species has prevailed for a very long time against its competitors is no guarantee that a rival will not emerge in the future and drive it into extinction.

The minimal content of the notion of 'adaptation' is simply the assumption that reality directs a selective pressure upon theories, as it does upon biological species. This metaphor of 'pressure' must in the final analysis be given a causal interpretation: reality contributes causally to the process through which theories are selected, and does so in systematic ways. The recognition of the influence of physical reality upon theories of natural science is compatible, of course, with recognition of the role of social factors, as well as the important fact that the fate of theories is not determined by their degree of fit with reality *tout court*, but by their relative degree of fit compared to that of rival theories. The winning theory is the one that fits nature better than its competitors. Here, there is a clear parallel with the fitness of living organisms, which is not determined simply by their degree of adaptedness to their inanimate surroundings, but by their level of adaptation relative to that of other organisms.

## 3. THE CONSTRUCTIVIST MODEL OF SCIENCE

Constructivists, on the other hand, tend to deny the platitude that reality plays a role in the formation of scientific beliefs about it. Indeed, I take this to be definitive of constructivism. Instead, constructivists hold that various social factors determine the content of theories of natural science. Here is Bruno Latour:

> For these parts of science [that is, the 'unsettled parts of technoscience', FC] our third rule of method will read: since the settlement of a controversy is *the cause* of Nature's representation not the consequence, we can never use the outcome – Nature – to explain how and why a controversy has been settled. (Latour 1987, p. 99).

And this is Harry Collins:

> The natural world has a small or non-existent role in the construction of scientific knowledge. (Collins 1981: 3)

This remarkable view comes in two strengths, associated with an ontological and an epistemological position, respectively.

The ontological position is to the effect that, rather than scientific theories being shaped by reality, it is really the other way around: scientific theories generate reality. Or rather: social reality determines our theories about natural reality, the latter theories in their turn determining natural reality. In this way, natural reality comes to be a reflection of social reality; it is as it were social reality projected out upon the rest of the universe. (Bruno Latour's constructivism is of the ontological kind, but differs from the one just suggested in that natural reality is not thought to be created by social fact but by a human 'praxis' which is somehow primordial to both social and natural reality (cf. Latour 1992; 1993). This difference is immaterial in the present context).

This doctrine should raise suspicions since social reality (or a Latourian 'praxis') is after all not ontologically autonomous: it is comprised of all sorts of things that have a physical aspect as well. This goes not only for such social hardware as houses, roads and machines, but even for human beings, who after all have physical bodies. How could there be a fixed social reality, or a social 'praxis', comprised of such things, prior to the fixation of physical reality which allegedly is something that happens along the way, in step with the development of physical theory? This appears entirely mysterious.

I have discussed these problems at length elsewhere, and shall not go into further details here (cf. Collin 1997). Suffice it to repeat my conclusion which is that ontological constructivism with respect to the physical world is untenable.

## 4. EPISTEMOLOGICAL CONSTRUCTIVISM

The other constructivist position is epistemological or methodological. Harry Collins's constructivism, as presented in the quote above, probably belongs to this type. It does not (necessarily) deny the existence of an independent reality, only its role in shaping scientific theorising. (But the two kinds of constructivism may be combined, as in Latour above.) There are many versions of this position, each inspired by different concerns. Most of them would exploit the distinction, not necessarily taken to be a sharp one, between sensory beliefs and theoretical ones. They would admit that the former are indeed caused by the world, but insist that the latter are not. The claim that theoretical beliefs are not caused by the world would apply in somewhat different ways within the context of discovery and within the context of justification. With respect to the former, the claim would be to the effect that theories of natural science are not generated by cognitive encounters with the world in the sense that human beings spontaneously generate the appropriate theoretical beliefs upon being presented with the appropriate sensory evidence. This point seems to be conceded on all sides, being inherent in the very distinction between observational and theoretical beliefs. Secondly, it would be argued that the world does not even play a role in the context of justification, or, better, of theory choice. To substantiate this position, an exacting notion of cause would be invoked, according to which the world could only be said to play a role in determining, and hence causing, the fate of our theories if its cumulative causal impact tended, over the course of time, to move science towards a specific end state, conceived of as a 'true representation' of the world. The critics would then add that the course of science over the last 350 years does not support the conclusion that it is approximating any such stable state (This is Putnam's notorious 'pessimistic induction', cf. Putnam 1978, p. 25).

The realist would be well advised to challenge the constructivist's definition of determination involved here as being far too strong. This definition would debar us from attributing any causal role to experience in determining the fate of theories even if there were clear cases of elimination of false theories through crucial experiments. Suppose that science proceeds in such eliminative fashion (that is, basically according to Popper's model of science), but that the series of eliminations fails to generate an approximation to any particular stable view. It would be very strange to conclude that experience plays no role in shaping theorising. It is indicative of the excessive strength of the constructivist requirement that it would indeed preclude us from attributing any causal role to environmental factors in biological evolution. For, notably, evolutionary pressure does not force any species, or life as such, in the direction of any particular state, such as optimal adaptation; it is not even clear what optimal adaptation would mean. There is no immanent end state to biological evolution. But it seems absurd to deny any determinative power to environmental factors on that account.

Barry Barnes and David Bloor have proposed an argument which is a variation on the same theme (Barnes and Bloor 1982, p. 34). They claim that the world's existence is a poor candidate to explain the nature of our theories about it, since the world is one, while the theories about it are many and highly diverse. They conclude that we cannot refer to the world to explain the variation, since we cannot explain variation by citing a stable factor.

The flaw in this argument is not hard to find. Between the world, which is one and the same, and our thinking about it, which is variable, a third, variable element is interposed, namely the amount and nature of our epistemic encounters with reality, in the form of perceptions, some of them the results of controlled experiment. These encounters produce an ever-increasing stock of experiences of the world, and it is these experiences that constitute the immediate cause of our theorising. This does not mean that the experiences generate theories but rather that they determine the fate of theories through the elimination of false ones.

Finally, a few words about Bruno Latour's argument as quoted above. This argument derives epistemological constructivism from ontological constructivism, citing the alleged fact of nature's being a product of the scientific research process in order to demonstrate that the research process cannot be shaped by nature in its turn: the determination cannot go both ways at once. This argument evidently has little appeal to those of us who do not subscribe to its ontological constructivist assumptions.

Still, it might be a little rash to dismiss the argument out of hand. It has an underlying point which may be saved even if we get rid of its ontological entanglements. The point might be rephrased thus: if we want to explain why one scientific hypothesis was finally chosen over a rival, it is no use pointing to the natural feature, S, which this hypothesis postulates as the explanatory factor. For the scientific community's recognition of the existence of S (although not the very *existence* of S, of course) only came by way of acceptance of the hypothesis H. The scientific community only came to believe in the existence of S as a consequence of accepting H; specifically, it did not accept H because it antecedently believed in the existence of S and hence would endorse a theory which captured that belief. Hence, we apparently cannot point to S to explain the fact of the acceptance of H.

However, this argument does not show the irrelevance of S, but only that reference to S must occur as part of a larger, two-tiered explanation of theory choice. It consists of a distal and a promixal part, of which the proximal part is normally the most salient. The proximal explanation might for instance take the form of an 'inference to the best explanation': Scientists are confronted with a set of observational phenomena, call it O. Among the hypotheses on offer, they judge that hypothesis $H_n$ provides the best explanation of O; hence they accept that hypothesis and hence the existence of feature S, which is a theoretical posit of $H_n$.

S itself enters the picture once we ask for an explanation of the observational phenomena O. The answer is that nature has a certain physical structure S which causes certain observational phenomena $O_1O_2O_3 ...O_n$ by its interaction with laboratory apparatus. These phenomena, in their turn, are recorded by scientists and lead to the adoption of a particular hypothesis $H_n$ on the grounds that this theory best explains those phenomena.

Of the two explanatory components, the proximal part will often hold the chief interest. This is the part which records the *grounds* for the choice, be they rational or irrational, correct or wrongheaded, explicit or implicit, or even deeply subconscious; and when we ask for an explanation of theory choice, we are normally interested in seeing the grounds for the choice. The distal account, on the other hand, is a purely causal one, couched in purely physical terms. It is, strictly speaking, no part of a sociological account of science at all; it is provided by natural scientists rather than sociologists. And it offers nothing by way of displaying the *grounds* for the choice of S: to say that the scientific community accepted the existence of S because S exists is completely uninformative in such a context. Still reference to S would have to figure in any distal account of why $H_n$ came to be accepted.

## 5. EXPLANATORY STRENGTH OF EVOLUTIONARY AND CONSTRUCTIVIST THEORIES

Let us next compare evolutionary and constructivist theories with respect to explanatory strength. We want our theories to be as strong as possible in the domain of explanation, *ceteris paribus*; and a crucial dimension of explanatory strength has to do with the range of phenomena explained. Constructivist theories might seem to have an advantage here, if only in the grandness of their aspirations: they claim to be able to explain the *genesis* of scientific hypotheses. Evolutionary theories, on the other hand, abnegate any explanatory ambitions as to the genesis of the fundamental items which compete for perpetuation. They only attempt to explain why, given that we start out with a field of competing items F, $F_i$ emerged as victorious in the competition. Let me stress that this claim is really definitive of the term 'evolutionary theory', as I use it here; it is not meant to give an accurate account of those theories which are normally labelled with that epithet. I introduce this *ad hoc* restriction on the term in order to produce a fruitful contrast between two kinds of theories, however labelled. (Historically, my use of the term may be close to Darwin's original version of evolutionary theory, before it was combined with Mendelian genetics to form the neo-Darwinian synthesis. But I shall bypass such historical issues here.)

In the framework of the sociology of science, an evolutionary theory in my minimalist sense would thus stake no claims within the 'context of discovery': it would abstain from trying to explain why particular hypotheses would pop up in the scientific community in the first place. It would only try to explain why, given a field of rival hypotheses, hypothesis $H_n$ won most adherents.

Certain constructivist sociologists of science, on the other hand, such as the authors of the Strong Programme, seem to be more ambitious. They want to explain the very genesis of theories. For instance, in *Scientific Knowledge and Socological Theory* (Barnes 1974), Barry Barnes argues energetically in favour of a determinist theory of science, specifically attempting to show how recognised cases of scientific creativity may be explained deterministically in sociological terms. I shall now try to show that this explanatory hubris brings trouble in its train.

## 6. PROBLEMS IN SOCIOLOGICAL EXPLANATION OF SCIENTIFIC THEORIES

Let us first briefly rehearse some basic points about the nature of explanation, according to the standard neo-positivist account (cf. Hempel 1965). An explanation is an argument where the event to be explained (the *explanandum*) follows from a number of premises (the *explanans*), either with deductive necessity or high statistical probability. The explanatory argument consists of at least one sentence of logically universal form, expressing a law of nature or an empirical generalisation, and one or more singular sentences expressing initial conditions, that is, particular features of the situation which render the protasis of the law applicable to it. In all but the simplest cases, the universal premise in the argument will be of very complex form, comprising an entire theory or even a cluster consisting of a main theory, surrounded by various auxiliary hypotheses. We need not record all this structure here. We can make do with a simplified model, consisting of the universal, theoretical component and the singular component(s). Thus we get the schema below:

Theory

Explanans

Initial conditions $I_1\ I_2\ I_3 \ldots\ I_n$

---

Explanandum

Now it follows from the tautological character of deductive logic, as well as of the calculus of probability, that no new information is allowed to appear in the conclusion of a valid argument. This extends to terms as well: any term that occurs in the conclusion of such an explanatory argument must somehow derive from the premises, and hence be found somewhere among them.[2] Given the structure of scientific explanation as just expounded, this means that the theoretical contents of the *explanandum* may derive from either of two sources: either from the general sociological theory, or from the premises spelling out the initial conditions (or from both of them – but here I shall concentrate on the pure cases which suffice to illustrate the point).

A little reflection shows that this fact presents the social constructivist with a dilemma: only a derivation of the former kind (call it Case I) can be said to explain the genesis of hypotheses in sociological terms. It does so, however, at the cost of presupposing an extreme logical and conceptual strength in the sociological theory involved; indeed, it is so immense as to constitute a *reductio ad absurdum* of Case I. The latter kind of derivation (Case II), where the theoretical and conceptual contents of the *explanandum* come from the initial conditions, is not saddled with such a huge inferential burden; on the other hand, it leaves the genesis of theories unexplained.

Let me illustrate these abstract claims with an example, starting with the kind of explanation where the theoretical contents of the inferred hypothesis (the *explanandum*) stem from the initial conditions (Case II). An interest-based theory about the selection of scientific hypotheses might assert that those scientific hypotheses are generally embraced, among the hypotheses on offer, which underpin the interests of the ruling classes; or that hypotheses are adopted differentially in different classes in proportion to their conformity with the interests of those classes. To explain any particular episode in the history of science, this theory would require additional information about the hypotheses circulating in the scientific community and about how this or that particular hypothesis would support the interests of the different classes. (This seems, for instance, to be the overall structure of Shapin's celebrated explanation of the differential endorsement of phrenology in Edinburgh, cf. Shapin 1975.)

This general theory carries a very light conceptual load; additional information, including additional concepts, are supplied by the premises spelling out 'initial conditions'. On the other hand, the theory does not explain the discovery, or creation, of the theories themselves, but only their reception, that is, their success or otherwise in the scientific community once they have been proposed. Shapin's analysis, for instance, does not explain why someone would come up with a theory linking the shape of the skull with the development of mental powers in the first place. (As a matter of fact, this happened not in Edinburgh but in Vienna through the efforts of one Franz Joseph Gall.)

In terms of our little simplified model, this case comes out as follows:

> Sociological theory:
> Among the scientific hypotheses on offer in any given situation, those will generally be embraced which underpin the interests of the ruling classes.
>
> Initial conditions:
> The following hypotheses were on offer: $H_1$, $H_2$ $H_3$ ... $H_n$.
>
> Among those hypotheses, $H_n$ conformed best to the interests of the ruling classes.
>
> ---
>
> Explanandum: $H_n$ was embraced by the ruling classes.

Let us now compare with the other alternative (Case I) where we use a theory that is strong enough to explain the very emergence of hypotheses. One kind of theory that would do the trick would be a functionalist one, to the effect that in any social situation, scientific hypotheses will emerge if their emergence and adoption are needed to satisfy a functional requirement of the society, or its ruling class. If we managed to render such a theory plausible, it could indeed be used to explain the emergence of a scientific hypothesis: It would explain by pointing to the conditions in society, or amongst its ruling classes, which are dependent upon the dissemination of precisely such doctrines for their perpetuation. (We may have general doubts about functionalist explanation, such as to the strength of the feedback mechanisms that are needed between the functional requirements and the items that satisfy them; but I shall bypass such worries here.)

In terms of our little simplified model, this case comes out as follows:

> Sociological theory:
> In any social situation, scientific hypotheses of the form $H_i$ will emerge if their adoption is needed in order to satisfy a functional requirement of the form $R_i$ in the society, or in its ruling class.
>
> Initial conditions:
> In situation S, there was a functional requirement $R_n$ of the society, or its ruling class.
>
> ---
>
> Explanandum: Hypothesis $H_n$ emerged.

Such a functionalist theory would have to have an immense inferential and conceptual strength, since it would have to be strong enough to imply, given additional information about the society in question, the basic features of any hypothesis which has actually been embraced by any significant scientific community. We may compare it to a computer program of such power that, when data about a society or community are fed into it, in particular data about its ruling class, the program generates a scientific hypothesis which is germane to the interests of that class. This is only possible if the functionalist theory in question contains, *inter alia*, all the terms of every hypothesis ever devised by mankind and subsequently adopted. Perhaps God could devise such a theory, but it is clearly way beyond the powers of man.

But it might be objected that the two contrasting examples do not exhaust the possibilities and that the apparent dilemma is a false one. It is easy to construct an example, it might be said, where the explanatory theory is not excessively strong, but still suffices to account for the genesis of a hypothesis. We use a functionalist theory as in the above example, but this time formulated without any intrinsic specification of the possible functional requirements, nor any intrinsic specification of the kinds of hypotheses whose dissemination would satisfy those requirements. The theory simply says that in any social situation, scientific hypotheses will emerge if their adoption is needed to satisfy a functional requirement in the society, or in its ruling class.

We next add a premise (initial condition) to the effect that adoption of a certain specified scientific hypothesis is needed to satisfy a functional requirement in the society, or in its ruling class. The hypothesis in question is specified in the abstract; no assumption is being made as to whether or not it has actually been proposed in the ongoing scientific debates. Still the functionalist theory now allows us to deduce that this hypothesis will at some stage be propounded by somebody, and will be embraced by the community in question.

Thus we apparently get the best of both worlds: we get explanatory power with respect to the genesis of a scientific hypothesis (moving within the context of discovery), but still avoid an excessive and forbidding inferential strength and conceptual load to bog down those functionalist theories and put them beyond human reach. Schematically, it would look like this:

> Sociological theory:
> In any social situation, scientific hypotheses will emerge if their emergence and adoption are needed in order to satisfy a functional requirement in the society, or in its ruling class.
>
> Initial conditions:
> In situation S, there was a functional requirement $R_n$ of the society, or its ruling class, which could only be satisfied by the emergence and adoption of hypothesis $H_n$.
>
> ---
>
> Explanandum: Hypothesis $H_n$ emerged and was adopted.

This explanation looks almost too good to be true; and indeed it is. Notice that the functionalist account makes use of information to the effect that the emergence of the hypothesis in question was needed in order to satisfy a functional requirement in the society, or in its ruling class (given the further features of the situation). Now this kind of information, which is here innocently tucked away as an initial condition, could only be supplied, as a general resource, by an auxiliary functionalist theory with the kind of extraordinary logical and conceptual strength which we tried to avoid in the original theory. (Or perhaps God could whisper it in our ears – but we should shun such theological shortcuts here.) If we want our sociological theory to be able to explain the advancement of science in general, the auxiliary component in our theory would have to be strong enough to generate, and compare, all the different scientific hypotheses that have actually been formulated, and in addition demonstrate that (the societal acceptance of) the particular theory under scrutiny was needed to satisfy a functional requirement in the society, or in its ruling class. The latter demonstration would necessarily involve the generation of a number of further, rival theories, formulated in the abstract, which might have emerged, but which would not have satisfied those requirements. Again, we are saddled with an immense burden.

Notice that information concerning the potential of a hypothesis to satisfy a functional requirement cannot be got cheaply from an analysis of the purely intellectual content of the doctrine in question. Such a purely hermeneutical, rationalistic procedure cannot tell us anything about the chances that such a theory will catch on in the general public, a parameter that is of crucial importance from a functionalist point of view. A doctrine to the effect that the aristocracy is a special race of superior beings, or that the king received his prerogatives directly from God, might perhaps supply an argument in favour of their interests in the abstract, but would be without practical value at a stage of historical development where such doctrines would be dismissed out of hand as naïve or ludicrous. Given the assumptions of constructivist sociology of science, the potential of a doctrine for gaining acceptance can only be determined on the basis of deep sociological theorising.

A final objection: the above argument might be thought to overlook the fact that reductive theories do not normally explain a phenomenon directly, but only indirectly, namely via *bridge laws*. For instance, the everyday term 'water' does not occur in the theories of physics or chemistry, hence no information about water may be inferred from them. On the other hand, the terms 'oxygen' (O) and 'hydrogen' (H) do so occur, and the theories in question thus suffice to explain the properties of a molecule consisting of an oxygen atom and two hydrogen atoms ($H_2O$). This has implications for water, once we add a bridge law to the effect that water is identical with $H_2O$. In a similar fashion, the laws of electromagnetics say nothing about light or colour, until we supply a bridge law stating that light is electromagnetic waves, and that differences in colour are differences in the wavelengths of light.

By parity of reasoning, a reductive sociology of science need not logically imply the hypotheses we want to explain. All that is required is that such implications ensue once we add relevant bridge laws, correlating theoretical terms in the sociological theory with corresponding terms in the *explanandum* hypotheses (in combination with the requisite initial conditions).

This objection is based upon a misunderstanding of the capacities of bridge laws, however. When a theory is linked to reality via a bridge law, the theory does not actually explain the occurrence of those features that are named only in the bridge law but not in the theory: it only explains such features as are encompassed by the theory. For instance, when we apply the theory of electromagnetism to the phenomenon of light, as licensed by the bridge law, 'light = electromagnetic radiation', we explain those features of light which it shares with electromagnetic radiation, such as that it is propagated in straight lines, that it moves with a speed in excess of 300000 kilometres per second, and so on. By contrast, we cannot explain why light of a certain wavelength presents a particular colour, such as green.

In a similar fashion, a sociological theory of scientific knowledge which operates via the mediation of bridge laws will only explain those features of its object which are named in those very theories themselves. Hence the charge stands that, in order for a sociological theory to explain the contents of theories of natural science, the sociological theory must somehow encompass the theories in question.

The way this explanatory dilemma is solved in practice by the sociologist of science is by lowering his or her explanatory aspirations. He or she does not really try to explain why, among all the hypotheses that mankind could possibly devise, *this* one was generated and adopted. Instead, he or she starts out with a limited number of hypotheses that are actually on offer, that is, that have actually been proposed, and proceeds to show why one of these would be favoured by the scientific community in the light of the latter's interests. But it is obvious that what we are offered now is an account of the *reception* of hypotheses already articulated, not an account of the *genesis* of hypotheses. The existence of a number of hypotheses is taken as a datum, as input to the explanation, and is not itself explained. They belong to the *explanans*, not the *explanandum*

## 7. ALTERNATIVE MODELS OF CONSTRUCTIVIST EXPLANATION

But perhaps this conclusion only comes about because we have been putting excessive demands upon our explanatory theories. We have assumed that an explanation only works if it makes the occurrence of the *explanandum* absolutely certain. But even the Hempelian model grants explanatory power to accounts which only make the *explanandum* highly likely. And even this modified Hempelian requirement represents a tightening of our everyday, and scientific, explanatory practice, where we often feel completely satisfied with accounts that not only do not imply the event to be explained with deductive necessity, but do not even assign any precise probability to it. For instance, we will explain a traffic accident by pointing to the oil slick which caused the car to skid on the curve. But the information provided will not allow us to deduce that an accident would necessarily occur, nor even bestow a specific probability upon that event. Still we will consider this kind of account to be adequate.

Moreover, the argument so far has moved within the general Hempelian framework for explanation. But this framework has been heavily contested in a debate over the nature of explanation that has been going on since around 1960. In that debate, a common charge against Hempel's model was its over-dependence upon examples from natural science; it loses its initial plausibility once we move to human affairs, whether on a societal macro-scale or on the micro-scale of individual action. Many alternative models have been proposed. Among these are various kinds of *intentional* explanation (purposive explanation, reason explanation, and so on).

The point is well taken in principle. It is true that the neo-positivist model of explanation developed by Hempel and others erred in insisting that explanations (explanatory arguments) make their conclusion certain, or at least highly probable. This requirement is only justified in the context of certain specific uses of explanatory information, but does not apply in general (for instance not to explanations of the 'how possible'-form). It is also true that alternative overall models of explanation exist, in particular with respect to human action, in which the explanatory power does not derive from the inferential strength of the tie between *explanans* and *explanandum*.

However, there is little comfort for the constructivist in these observations, given the strategic situation in which they find themselves vis-à-vis the traditional, rationalist picture of science. The strategy adopted by constructivist sociologists of science has been to insist that their rationalist opponents produce a perfectly stringent standard of rational theory choice in terms of which to explain the development of science. There has been no patience with vague formulations. Not getting what they asked from the rationalists, the constructivists have been quick to conclude that rationalist accounts are bunk, and to declare sociological explanations to be the only viable alternative. (As it happens, the rationalists themselves have looked for the same kind of mathematically exact standards. This is why some constructivists look suspiciously like frustrated rationalists.) The philosophers' vague hand-waving in the direction of 'standards of rationality' is now to be replaced by the sociologists' precise and no-

nonsense explanations in terms of interests and so forth. Now it clearly undermines this strategy if the constructivists are incapable of delivering explanations that are superior to the rationalists' in strength. A vague account of an episode in the history of science, referring to various 'interests' without spelling out precisely what those interests are, and why they are better served by the hypothesis in question than by an indefinite number of different ones, is hardly better than explaining the same events by referring to the 'rationality' of the choice, but leaving the standards of rationality conveniently indeterminate.

This objection applies in particular to the proposal that intentional explanations should replace the Hempelian kind. For it would seem that if intentional explanations are to be made scientifically precise, to serve as a tool for the sociology of science, they would inevitably move in the direction of rational, decision-theoretic accounts (rational choice theory). We would now be urged to see the choice of scientific hypotheses as being the result of a careful calculation of benefits, according to some calculus of expected utility, or whatever. This kind of theory is utterly rationalistic, and its adoption would go entirely against the spirit of the new sociology of science with its scorn for rationalistic models of theory choice. The only change would be that theoretical rationality had now been replaced by the practical variety as the medium of theory choice. (For a related criticism, see Fuller 1993, Chapter 4.)

We may note in passing that in the text I have used as an example, Barnes's *Scientific Knowledge and Sociological Theory*, the author explicitly commits himself to a strictly deterministic account and hence, by implication, to the most stringent version of the Hempelian analysis of explanation. Hence no injustice is done to Barnes in saddling him with the implications of a strict Hempelian view. In fairness to Barnes, it should be noted that in the text in question, he grants reality a role in the shaping of science. This role is apparently exercised in the *selection* of hypotheses, whereas the social determinist view pertains to the genesis of same. The above argument was precisely directed against the latter kind of view.

## 8. THE PROBLEM OF RELATIVISM

Towards the end of the last section, however, we digressed from our original topic, which was the comparison of evolutionary and constructivist theorising, and strayed into a comparison of constructivist and traditional rationalist theorising instead. Let us return to our original discussion.

Social constructivists might accept the upshot of the above discussion, namely that their efforts do not suffice to explain the genesis of scientific hypotheses, and thus that they do not do any better than evolutionary theories on this score. Still they would insist that their theories are more powerful than evolutionary theories, since they can explain certain features of theorising on which evolutionary theories must necessarily be silent, namely concerning the nature of the factors determining the reception of theories in the scientific community. Evolutionary theories, it would be alleged, cannot tell us anything on this score. The reason is that the only answer that evolutionary theories could give would be to the effect that the scientific community adopts the theory which strikes it as the best, given the evidence. *Why* the theory strikes it as the best, in other words why the community subscribes to the standards of scientific merit that it does, lies outside the purview of evolutionary theory. Hence, social constructivist theories provide a better and more complete account of the development of science than evolutionary theories.

I think it is true that constructivist theories offer us the potential promise of going beyond evolutionary theories of the growth of science on this point. Yet great caution has to be shown in describing this advance so as not to weaken the constructivist position rather than strengthening it. Moreover, I believe that when a proper description has been found, that description will be seen to point beyond itself and towards a better way to transcend the limitations of evolutionary theory.

Let us first look at the dangers inherent in the step which the constructivists take beyond evolutionary theory. The constructivists purport to explain the standards which dictate the choices between competing theories, and to explain them in terms of familiar sociological categories. The sociological account will talk about *interests*, *power* or *ideology* where the traditional philosophical account talks about rationality. Sometimes, the replacement of rationality with interests or power forms part of an effort to debunk the pretensions of (natural) science: constructivists often purport to show that so-called scientific rationality is a mere rhetorical facade designed to hide the grim face of interest. However, as critics have been quick to point out, this strategy will immediately backfire. The social constructivists' position is itself based upon scientific evidence; indeed, constructivists pride themselves on having replaced vague philosophical speculation about the nature of science with hard-nosed empirical investigation. But if scientific rationality in general is declared to be bunk, and merely a facade for underlying interests, the constructivists undermine their own position: that, too, will now merely be the facade of underlying interests or ideology (cf. Laudan 1990, Chapter 6). The only way to deflect this criticism, within the framework of a debunking version of the sociology of knowledge, is to insist that there is a crucial difference between natural and social science on this score. In other words, the constructivists must maintain that the theories of natural science are social constructions, reflecting features of science as a social process rather than features of the natural world, but that social science, including the sociology of knowledge, offers a true representation of social reality. Claims along these lines have been aired by Collins (Collins 1992, pp. 187–9) and by Fuller (Fuller 1993, p. xiv). But it must be very hard to show why those doubts which have been raised concerning the rationality of natural science would apply with less force to social science. After all, those doubts spring from general arguments, and are not tied to any particular branch of science at all.

A much superior strategy on the part of constructivists is to claim that what emerges from the work of the sociology of science is not the discreditation of rationality, but instead its *relativisation*. In linking standards of theory choice to certain interests or ideologies in the community, it is not being implied that those standards are somehow invalid but merely that their validity is linked with, and indeed relative to, certain fundamental parameters of the community and its dominant social practices. This position is not self-defeating. The constructivists may now without worry apply their principles to their own doctrine, admitting that the standards of (scientific) rationality adopted in their sociological work are no doubt relative to the community and the epoch in which they are used. They will only point out, in mitigation, that they conform to the standards that everybody else abides by in contemporary social science.

This position has been most fully developed in the Strong Programme of David Bloor and Barry Barnes (cf. Bloor 1976). They insist that any programme in the sociology of science must be reflexive, or self-applicable, which rules out the debunking stance dismissed above. They favour a programme of empirical relativism, which has its philosophical basis in Wittgenstein's philosophy of language-games and ways of life. Scientific communities represent different practices and different language-games. There is no higher and independent point of view in terms of which to assess these. The result is a relativist position.

This position, however, will only be acceptable if it can be developed so as to handle the following difficulty. Divergent standards of cognitive rationality will lead to divergent theories of the world. Now whenever we have two competing views of the world, we know that at least one of them is false. Hence, the cognitive determination of thinking by social factors compels at least one of the two competing schools to embrace a falsehood. This must inevitably generate a desire to break this determination. We are forced to seek a higher epistemic ground from which the alternatives can be surveyed without prejudice. Hence a sociology of science conducted according to this paradigm would tend to be self-defeating: its application to reality would tend to create a process which would render it inapplicable.

A possible way out for the constructivists might be to deny that the divergent accounts of reality are really incompatible, and hence their opposition something that needs to be overcome. Instead, those divergent frameworks can be assimilated through 'relationism' or 'perspectivism', which has it that there is no unique truth about the world, but just as many truths as there are points of view from which the world is surveyed and thought about. In particular, each society or social group subscribes to a conception which is determined by its special cognitive access to reality, its special view upon reality, in a generalised sense of 'view' which is comprised of a theoretical and conceptual aspect as well.[3]

Unfortunately, there are problems in applying this epistemological model to the present case. Relationism is an attempt to render logically innocuous the fact that different social groups often embrace divergent conceptions of reality. The technique is to make the contents of those conceptions relative to some parameter which varies as between social groups. In this way, we avoid a straightforward contradiction between those positions. Such a reinterpretive strategy is sometimes legitimate. A good example is the theory of relativity, which has taught us that facts about the simultaneity of physical events are relative to the state of motion of the systems of reference within which the events are described. Observers located in two different systems of reference moving in relation to each other will report the temporal relationships between events differently, but the divergences are systematic and predictable. By making this dependency explicit, in other words by adding a reference to the framework of observation as an index on temporal predicates, we avoid statements about simultaneity in different frameworks of observation emerging as stark contradictions.

The crucial point here is that 'observer' is not to be taken in a subjective or cognitive sense: the observations may be recorded in perfectly objective media, such as light spots on photographic plates linked to clocks. This means that in the theory of relativity, it is possible to construct a consistent objective picture of reality, namely a picture of reality independent of our cognitive relation to it. The picture is objective, but not absolute: it is relative, but to a feature that is not in itself cognitive, but an objective one, namely the state of motion of the observer.

Let us now compare this with a hypothetical example in which relationism is invoked to explain disagreement motivated by different interests. Take the classical episode of Galileo arguing the case of his 'New system of the world' against the Catholic clergymen. Galileo claims to be able to see craters on the moon, the bishops insist that they are not there to be seen. The interests of the bishops are obvious: they wish to uphold the world view endorsed by the Catholic church, which involves subscribing to the Aristotelian world picture with its doctrine of perfect circular motion, and so forth. (What Galileo's interest is I will leave to social constructivists to worry about.) Now in this case there is no such thing as giving an objective but relational account of the world that will save consistency. We cannot describe how the world objectively is – that is, in terms of features which could be recorded by such devices as cameras and clocks – while at the same time the features in question vary as between Galileo and the clergymen. What we can give instead is an account of what the world is like, in objective and *absolute* terms, and why the sensory data generated by it would tend to induce in Galileo the belief that the moon is pitted with craters, while the same sensory information would generate the opposite belief in the clergymen. But this is evidently not a reconciliation of the parties' astronomical beliefs, using the strategy of relationism, but merely an explanation of why their beliefs differ. No picture of the world has been provided which is at once objective – that is, referring to features of the world, not of the observers – and at the same time relational, that is, couched in terms of features that vary as between observers.

We may use Shapin's account of the reception of phrenology in Edinburgh in the early 19th century as another example. (It is not clear whether Shapin would subscribe to relationism. I impose that reading upon his article here.) Let us accept Shapin's point that the working class's enthusiasm for phrenology was caused by the latter's potential for favouring the interests of this stratum of society, whereas its rejection by the middle and higher classes was motivated by their contrary interests. In this case again, there is no parallel to the situation in relativity theory: no account could be given of the relationship between cranial shape and the development of mental powers which would be at once objective – that is, which would record features of the world, not of somebody's beliefs about it or attitudes to it – but would still be perspectival in the sense of varying between the classes in question. The stances of the working class and the middle and higher classes with respect to phrenology remain in direct and stark contradiction.

## 9. CONCLUSION

I conclude that the relationist model of knowledge is not applicable to, and hence not capable of dissolving, the kind of disagreement between different scientific orientations which allegedly springs from divergent background interests. Such a disagreement remains a naked contradiction with respect to one and the same reality, resistant to elimination by invocation of the conception of different aspects of one and the same world. This means that such disagreements must force us to develop a neutral epistemic platform from which the contradictory claims may be adjudicated. This is so in particular if we come to possess precise deterministic theories linking the acceptance of scientific hypotheses to determinate social factors. Given that science is essentially an intersubjective, collaborative enterprise which proceeds by establishing consensus among scientists, science as an institution cannot live with a situation where it looks as if scientists of different backgrounds are destined to disagree, and disagree irreconcilably. Science must inevitably construe such disagreements as indicating the presence of factors interfering with the proper conduct of science, and must use the very theory which raises this spectrum as a tool for changing the scientific practice and neutralising such influences. Such disagreements will inevitably force us to seek for a neutral point of view that would dissolve them. True, it would not necessarily enable us to find it. But it might help us in identifying those features in two rival theories which appealed specifically and differentially to their respective protagonists. This might eventually lead to ways of assessing the merits of those theories that would not trigger those interests, such as by using experts who had no vested interest on either side. It would seem that the more precisely the sociology of science could identify the factors that create dissensus, the easier it would be to neutralise them. Another way to create consensus might be to look for evidence which both parties would agree to accept as an *experimentum crucis* with respect to the issue under debate.

Let me emphasise that the above is not just a normative conclusion, advising science about what it ought to do in the event of a certain kind of theory emerging on the sociology of science scene. It is a prediction, and I would say a fairly safe one, as to what would actually happen if such theories became available. In our modern world of Big Science, the organisation and governance of science is a highly professonalised, highly reflexive enterprise. Information about how science works and how it could be improved upon is eagerly sought by science administrators at all levels and when found is quickly fed back into the practice of science management. There is no doubt that, if the sociology of science produced theories demonstrating a precise linkage between partisan societal interests and particular standards of hypothesis assessment, science managers would try to design institutional frameworks and organisational mechanisms with which to overcome such partisanship.

Thus we need to transcend evolutionary models of science in order to account for the societal aspects of the process of theory choice. But it turns out that social constructivist models are not the right way to achieve this. The above reflections show that both evolutionary and social constructivist theories are likely to be superseded by other kinds of theory, once reliable empirical knowledge about science begins to accumulate. Science is a process of learning or self-education where the results of studying science with scientific methods will themselves be fed back into the scientific process, modifying and improving the latter. Hence, what we need in order to understand science is some kind of self-reflexive model which does not merely accept existing scientific practices as sacrosanct, but adopts a critical stance. Steps towards building such a model have been taken in Steve Fuller's 'social epistemology' which combines a sociological and a critical stance (Fuller 1993). But these are only the opening moves in a process which still has a very long way to go, and huge issues remain to be resolved concerning, for instance, the extent to which the criticism of existing scientific practices can only be immanent, that is to say, in terms of standards intrinsic to those very practices, and to what extent criticism can be in terms of external standards. The answer to such questions can only emerge from future work.[4]

## NOTES

1 Cynics might suggest that those features of biological organisms which are normally explained by reference to 'sexual selection', such as the deer's antlers and the peacock's tail which appear to have no survival value, make a good parallel to those features of science which appeal specifically to the tastes of the 'consumers' of such theories but have little to do with their fit with reality. They might be such features as the 'beauty' of a theory, or its 'simplicity'; interesting theories are often colloquially even referred to as 'sexy'. But I think this analogy has a very limited scope.

2 There is a trivial exception to this principle which should not bother us: If a sentence P follows from a set of sentences X, then so does the disjunctive expansion of P – P or Q – where Q and its component terms need not occur in X. It is obvious that such expansion does not serve to explain Q, and does not introduce its terms in a way that is scientifically legitimate. Hence we may ignore it.

3 Notice that this position differs from the form of ontological constructivism we examined initially in not claiming that there are multiple realities, and a *fortiori* in not claiming that such multiple realities are generated by the way we think about them. All that is asserted is that there exist multiple different points of view, in an extended sense, upon one and the same reality.

4 I am grateful to the participants in the workshop on 'Science as a Spontaneous Order' in Skagen in January 2000 for their responses to my oral presentation of the present material. Thanks are also due to Claus Emmeche for his comments on a draft version of the material, and to the editors for their suggestions for improvements.

## REFERENCES

Barnes, Barry (1974), *Scientific Knowledge and Sociological Theory*, London: Routledge & Kegan Paul.

Barnes, Barry and David Bloor (1982), 'Relativism, Rationalism and the Sociology of Knowledge', in Martin Hollis and Steven Lukes (eds), *Rationality and Relativism*, Oxford: Basil Blackwell, pp. 21–47.

Bloor, David (1976), *Knowledge and Social Imagery*, London: Routledge & Kegan Paul.
Collin, Finn (1997), *Social Reality*, London: Routledge.
Collins, H. M. (1981), 'Introduction: Stages in the Empirical Programme of Relativism', *Social Studies of Science*, **11** (1): 3–10.
Collins, Harry S. (1992), *Changing Order*, Second edition, Chicago, IL: University of Chicago Press.
Fuller, Steve (1993), *Philosophy of Science and its Discontents*, Second edition, New York, NY: The Guilford Press.
Hempel, Carl G. (1965), *Aspects of Scientific Explanation – and Other Essays in the Philosophy of Science*, New York, NY: The Free Press.
Latour, Bruno (1987), *Science in Action*, Milton Keynes: Open University Press.
Latour, Bruno (1992), 'One More Turn After The Social Turn', in E. McMullin (ed.), *The Social Dimensions of Science*, South Bend, IN: University of Notre Dame Press, pp. 272–94.
Latour, Bruno (1993), *We Have Never Been Modern*, Cambridge, MA: Harvard University Press.
Laudan, Larry (1990), *Science and Relativism*, Chicago, IL: University of Chicago Press.
Popper, Karl R. (1972), *Objective Knowledge*, Oxford: Clarendon Press.
Popper, Karl R. (1974), *Autobiographical Sketch,* in Paul Arthur Schilpp (ed.), *The Philosophy of Karl Popper*, La Salle, IL: Open Court.
Putnam, Hilary (1978), *Meaning and the Moral Sciences*, London: Routledge & Kegan Paul.
Radnitzky, Gerard and William W. Bartley III (eds) (1987), *Evolutionary Epistemology, Rationality, and the Sociology of Knowledge*, La Salle, IL: Open Court.
Shapin, Steven (1975), 'Phrenological Knowledge and the Social Structure of Early Nineteenth-Century Edinburgh', *Annals of Science,* **32**: 219–43.
van Fraassen, Bas C. (1980), *The Scientific Image*, Oxford: Clarendon Press.

# 5. A neo-Darwinian model of science

**Thorbjørn Knudsen**

## 1. INTRODUCTION[1]

Darwin is usually viewed as the isolated genius working alone, a view supported by the fact that early on he apparently told no one about evolution or natural selection. Later, he broached the topic of evolution with a very few, and even later he only told a few about natural selection. Yet, as shown in the following, a recollection of the delayed publication of *Origin* indicates that even if Darwin did much of his work in isolation, face-to-face interaction with a tightly knit group of friends and colleagues played an important part in establishing and further developing his ideas. Thus I use Darwin, often perceived as the isolated genius, to illustrate that social interaction and possibly implicit knowledge acquired in face-to-face groups plays a crucial role in forming the scientist's presumptions and conceptual categories. As described in the following, Darwin's social activities somehow helped develop the implicit knowledge that allowed him to benefit from reading and experimentation. How did this happen?

The attempt to answer this question has inspired development of the neo-Darwinian model of science proposed here. It is a nested model whose inheritance track is constituted by implicit and explicit knowledge, operating in two (ontologically) distinct but connected trails.[2] This model is based on what I refer to as local emulative selection, the unconscious selection (emulation) of knowledge that takes place in social interaction.[3] An additional inspiration for the proposed model is that also in *Origin*, Darwin had described the notion of unconscious selection in terms consistent with the process of local emulative selection proposed here.

The purpose of the present essay is to further develop previous evolutionary models of science by adopting a nested selection-retention structure whose foundation is the selection of implicit knowledge. It is organised as follows. The section 'Darwin's Delay' argues that a possible reason for Darwin's delay in publication of *Origin* is that he needed time to acquire the implicit knowledge necessary to further develop his insights and to convey his results to the scientific community. This introduces what I refer to as the support-hypothesis, which establishes the basis for the neo-Darwinian science model proposed here. According to the support-hypothesis, explicit knowledge must be supported by an underlying structure constituted by implicit or tacit knowledge. The section 'Darwin's Difficulty' illustrates the support-hypothesis by pointing out that a crucial reason for Darwin's delay was associated with the difficulty in changing the conceptions of inheritance he had acquired early on.[4] It is further pointed out that a crucial problem for Darwin, and a problem that still plagues most evolutionary models of science, was the conflation of the genetic and the somatic trail. With the benefit of this insight and over a hundred years' evolution of evolutionary thought, I then introduce some terminology and a general understanding of evolutionary explanations. Having briefly reviewed a small sample of prominent evolutionary models of science in the section 'Evolutionary Models of Science', the ensuing section 'A neo-Darwinian Model of Science Based on Local Emulative Selection' introduces the proposed neo-Darwinian model of science.

## 2. DARWIN'S DELAY

This section introduces the support-hypothesis, according to which explicit codified knowledge must always be supported by uncodified implicit knowledge, and considers Darwin's delay in publication of *Origin* as an illustration of this idea. Although many reasons have previously been considered as causes for Darwin's delay, it is argued that the support-hypothesis can be viewed as an additional reason. It is thus argued that a possible reason for Darwin's delay is that he needed time to acquire the implicit knowledge necessary to further develop his insights and to convey his results to the scientific community. That is, Darwin needed time to develop the necessary implicit knowledge that could support his explicit selection principle. The significance of the support-hypothesis is to establish the basis for the neo-Darwinian science model proposed in this chapter.

Having read Thomas Malthus's *Essay on Population* on 28 September 1838, Darwin experienced the revelation that allowed him to formulate his theory of natural selection as an explicit hypothesis. By following Darwin's train of thought throughout his notebooks, one can see that on previous occasions Darwin had recognised the principle of selection in nature, and had even momentarily recognised the transforming power of chance adaptations, which lies at the core of his selection principle (Richards 1987). Yet this fleeting awareness remained implicit knowledge for Darwin until 28 September 1838 (ibid.). Although Darwin recognised that in domestic breeding, the breeder could alter species, he would first draw the analogy with nature when he discovered a causal force that could drive nature in the same manner as the desire for novel variety drove the breeder:

> I soon perceived that selection was the keystone of man's success in making useful races of animals and plants. But how selection could be applied to organisms living in a state of nature remained for some time a mystery to me.
>
> In October 1838, that is, fifteen months after I had begun my systematic enquiry, I happened to read for amusement Malthus on Population, and being well prepared to appreciate the struggle for existence which everywhere goes on from long continued observation of the habits of animals and plants, it at once struck me that under these circumstances favourable variations would tend to be preserved, and unfavourable ones to be destroyed. The result of this would be the formation of new species. Here, then, I had at last got a theory by which to work.

In this passage, an excerpt from Darwin's *Autobiography* ([1887] 1969) quoted in Richards (1987, pp. 98–9), we are offered a recollection of the conception of one of the most famous scientific principles. In Richards's account of the evolution of Darwin's thought, the focusing event of September 1838 must be seen against the background of the ideas he had previously established (see also Schweber 1977a). Darwin had on many occasions ruminated on the principle of selection and yet failed to see its significance as a general causal mechanism that could drive species change also in the wild. And when he finally formulated his selection hypothesis in explicit form, he continued to extend the reach and depth of his discovery over a period of twenty years before he published the results in *Origin* (see for example Richards 1987). I believe that Darwin's own account of the explicit formulation of his selection principle does not reflect the mere conversion of implicit to explicit knowledge. Something else was involved: for Darwin, prior implicit knowledge acted as a seed that gave rise to an explicit principle, which could subsequently be reflected upon, communicated and further developed. In Darwin's own description of his delay in the introduction to *Origin*:

> On my return home [from the *Beagle* voyage], it occurred to me that something might perhaps be made out of this question [the origin of species] by patiently accumulating and reflecting on all sorts of facts which could possibly have a bearing on it. After five years' of work I allowed myself to speculate on the subject, and drew up some short notes; these I enlarged in 1844 into a sketch of the conclusions, which then seemed to me probable: from that period to the present day I have steadily pursued the same object. I hope that I may be excused for entering on these personal details, as I give them to show that I have not been hasty in coming to a decision. (Darwin [1859] 1985, p. 65)

It was not as if implicit knowledge had suddenly been converted to an explicit form that could readily be understood by Darwin and then immediately communicated to the scientific community. More than 20 years elapsed between his explicit formulation of the principle of natural selection in 1838 and his publication of this result in *Origin* in 1859. Normally it is thought that fear of the consequences of shattering contemporary world views held Darwin back until he received Alfred Russel Wallace's essay that shocked him by expressing a selection theory that read as his own.[5] Although concern for the reception of his theory undoubtedly played a role in delaying Darwin, Richards (1987) points to an additional convincing and perhaps even more important reason. Darwin simply needed time to better understand the reach and depth of his own selection principle and in particular to work out how it could overcome the possible difficulties to be raised against it. As I see it, having formulated the selection principle in explicit form, Darwin needed time to develop the implicit knowledge necessary to form the conceptual categories that served to extend his new insight. That is, to understand and communicate the first explicit formulation of the selection principle, Darwin needed to develop additional implicit knowledge.

Thus Darwin seems a perfect example of what I refer to as the support-hypothesis: explicit codified knowledge must always be supported by uncodified implicit knowledge. This support-hypothesis can be derived as the common overlap among a number of distinct streams of literature: experimental psychology (Reber 1993); the psychology of professionalisation (Sternberg and Horvath 1999); the psychology of human error (Reason 1990), and philosophy (Polanyi 1957; 1966; Ryle 1971; Wittgenstein 1969).[6] Also, theories of choice in organisations (March and Simon 1958; Nelson and Winter 1982) have emphasised how implicit uncodified knowledge plays a role in the formation of skills, routines and semi-automatic behavioural programmes. Recently, there has been a renewed interest in the properties of implicit knowledge and the interaction between implicit and explicit forms of knowledge (Nonaka and Takeuchi 1995). These recent contributions view invention and innovation as somehow overcoming the difficulty in converting implicit forms of knowledge into explicit knowledge. By contrast, the support-hypothesis offered in the present essay is based on the idea that implicit and explicit knowledge are distinct forms of knowledge whose combination constitutes a process of knowledge accumulation, with implicit knowledge being necessary for the expression and absorption of explicit knowledge.

A number of reasons for Darwin's explicit formulation of his principle of selection have been pointed out. There is no doubt that Malthus provided the spark, but an additional reason is that Darwin had previously developed the conceptual categories required to appreciate what he saw in Malthus, and subsequently to express his insight in an explicit form that eventually could be appreciated by his colleagues and friends. Darwin had gradually developed the required conceptual categories through extensive reading of the Scots moral philosophers and frequent face-to-face interactions with prominent scientists in the period after he returned from the *Beagle* voyage (Richards 1987). I am not suggesting Darwin knew what he would need in order to appreciate Malthus. On the contrary, what I am proposing is that the conceptual categories Darwin had happened to develop not only allowed him to formulate his selection principle in explicit form but also to do so in a manner that potentially made it understandable and interesting to his friends and colleagues. The process of formulating his principle in explicit form is therefore entwined with the requirement for its further development, that it must be conceived in terms that could make sense to the scientific community of his time.

There is a further point to be considered. Darwin did not merely acquire explicitly formulated ideas by reading them in books and journal articles. The conceptual categories that allowed him to perceive what he read evolved through face-to-face interaction and written correspondence with a tightly-knit network of influential and closely connected colleagues and friends. After the *Beagle* had docked on 4 October 1836, Darwin spent a brief time at his father's house in Shrewsbury and then moved temporarily to Cambridge in mid-December 1836 (Richards 1987). Later, in early March 1837, he settled in London where he remained until 1842 (Richards 1987). In this period he became intimate with leading members of the scientific establishment:

> ... dining with the Lyells, renewing friendship with Whewell, Sedgwick, and Grant, and attending meetings at the Geological Society, the Zoological Society, and the Royal Society. ... Yet within this public sphere of an extraordinarily active scientific life, he simultaneously inhabited a more private intellectual environment, in which he worked to develop a new theory of species change. The various features of the public environment – the problems, the literature, the strategies of argument, and the zeal for scientific fame – these infiltrated the cognitive space of his emerging species theory, and thus provided important elements of that intellectual ecology against which his theoretical conceptions evolved. It would be a long time, however, before, the ideas nurtured in his private notebooks and essays would invade the expansive terrain of Victorian and scientific life. (Richards 1987, pp. 83–4)

As a first step in establishing the evolutionary significance of implicit knowledge for the expression of explicit scientific knowledge, it is important to consider how the two forms of knowledge are acquired. As the quote from Richards (1987) suggests, the public environment provided both the explicit and implicit knowledge that fed into Darwin's evolving theoretical conceptions. Explicit scientific knowledge was codified in the books and journals of science and could simply be acquired by reading. Implicit knowledge, by contrast, was conveyed to Darwin through his extensive social interaction with the scientific community.

It is interesting to note that Darwin seems to have acquired a genuine taste for reading rather late. Neither public school nor the medical studies he pursued at the University of Edinburgh inspired Darwin's intellectual appetite (Richards 1987): he thought the studies were dull, useless and dreadful, and agreed to his father's plan that he should complete an education at Cambridge and then take religious orders (Richards 1987). Darwin's father simply thought his son a very ordinary boy below common standard intellect and found this would preclude any professional life (Darwin's *Autobiography,* quoted in Richards 1987). Although Darwin, according to Richards (1987), seems to have found the courses at Cambridge little more interesting than the course work at Edinburgh, he was attracted to, and continued to study, zoology and geology. Already, at the University of Edinburgh, Darwin had strayed into the natural sciences, had read Jean Baptiste de Lamarck (1774–1829) as well as his grandfather Erasmus Darwin's (1731–1802) *Zoonomia*, enrolled in geological and zoological lectures and continued to collect beetles, an interest he had acquired in his boyhood (Darwin's *Autobiography,* quoted in Richards 1987). Later, at Cambridge, he read *Evidences of Christianity, Moral and Political Philosophy* and *Natural Theology* by William Paley (1743–1805).[7] Whereas *Evidences* introduced Darwin to the question of morality in Christian theology, *Natural Theology* was an excellent introduction into natural history (Mayr 1982). Also, Alexander von Humboldt's *Personal Narrative* and John F. W. Herschel's *Introduction to the Study of Natural Philosophy*, which he read at his last year at Cambridge, had a lasting influence on Darwin's career. Herschel's study, in addition to providing a solid exposition of scientific methodology, further fuelled Darwin's ambition to become a natural scientist; and von Humboldt's *Personal Narrative* directed Darwin's ambition towards becoming an explorer (Darwin's *Autobiography,* quoted in Mayr 1982).

The point is that Darwin directly absorbed scientific knowledge and personal inspiration in explicit codified form. Thus, during the voyage of the *Beagle* (1831–1836), Darwin read the two volumes of Charles Lyell's (1797–1875) *Principles of Geology*, which had a lasting influence on his thinking, including his conception of Lamarck (Mayr 1982; Richards 1987). Darwin actually claimed he was 'self-taught'. According to Mayr (1982), this was not an unjustified claim since Darwin got his real education from observing and reading. As Darwin himself noted, 'There are no advantages and many disadvantages in lectures compared with reading' (Darwin's *Autobiography,* p. 47, quoted in Mayr 1982, p. 397). In consequence Mayr (1982, p. 397) observes:

> To mention the books that impressed him [Darwin] as a young man is, therefore, as important or more so than mentioning the professors whose lectures he attended in Edinburgh and Cambridge.

Using clear evidence from Darwin's *Autobiography*, Mayr (1982) and Richards (1987) agree on the impression of Darwin as a 'self-taught' man who directly absorbed scientific knowledge in explicit codified form. And that could be the end of the support-hypothesis, that explicit codified knowledge must always be supported by uncodified implicit knowledge. There seems to be agreement that Darwin on his own read Paley, and it is clear that he read Lyell in isolation. On top of that, he 'happened to read for amusement Malthus on *Population'* (Darwin's *Autobiography*, pp. 119–20, quoted in Richards 1987, pp. 98–9). Note that Darwin absorbed the explicit codified message in Malthus. That seems to be the final refuting evidence against the support-hypothesis and in favour of the alternative hypothesis, that explicit and codified scientific knowledge can be directly absorbed and utilised by its recipient. And yet, we can ask, what was it that enabled Darwin to grasp implications of Malthus's argument that reached beyond those Malthus himself considered? And why Darwin's delay in conveying his insight to the scientific community?

As I will argue in the following, the answer to both questions is indeed consistent with the support-hypothesis. Darwin was able to appreciate in Malthus what none before him had seen, partly because he had acquired the necessary preconceptions by face-to-face interaction with a large number of the leading scientists of his day; preconceptions he continued to develop and qualify by systematic gathering of empirical evidence during the *Beagle* voyage and throughout his life.

But why the delay? According to Richards (1987), Darwin's first nebulous thoughts on selection theory took a much clearer shape after the Malthusian revelation. Still, it was not as if the theory of natural selection just presented itself in clear form that could readily be conveyed to the scientific community. Richards (1987) presents a detailed and convincing history that Darwin needed a long period of time to grasp the reach of the selection principle, which had presented itself to him in 1838. In addition to the fear of the public reception of the 'continuity claim', that there was continuity rather than divide between man and beast, Richards (1987) points to the following reasons for Darwin's inertia. First, as explained below, he simply required time to find the solution to a crucial problem associated with instinct inheritance. Second, he also required time to develop an evolutionary theory of morality, a problem he first addressed head-on in *Descent*, published twelve years after *Origin*. Third, Darwin had a desire to convince his friends among the natural theologians, Charles Lyell, Joseph Hooker, and John Henslow, that natural selection of instincts was a reasonable possibility.[8] Fourth, there is the more pragmatic reason associated with the need for Darwin to establish his reputation in the scientific community, a goal he pursued through extensive publication between 1839 and 1856. Finally, anxiety over the negative scientific reaction to an evolutionary theory published by Robert Chambers in 1844 may have cooled Darwin's taste for immediate publication.[9]

Richards (1987) rightly points out that the error usually committed has been to consider one factor in isolation as the explanation for Darwin's delay. Accepting this warning this chapter advocates that the support-hypothesis, which includes some aspects of the first three items in the above list, must be considered as an additional reason for Darwin's delay. According to the support-hypothesis, Darwin needed time to acquire the implicit knowledge necessary to further develop his insights. Without refined conceptions, Darwin could not overcome the difficulty for the selection argument that was posed by the scientific community.

The difficulty Darwin had to overcome was not an imaginary problem, but rather the difficulty in overcoming the objections he *knew* would be raised by colleagues and friends. Both in order to convey his insight so that his colleagues could appreciate it, and in order to foresee the possible objections raised against it, Darwin had to share conceptual categories with the relevant scientific community. But how did this happen? As explained further in the following section, part of the scientific community was, for Darwin, his face-to-face group, consisting of close friends and colleagues: people he frequently interacted with. One possible answer to the crucial question of how Darwin had acquired the implicit knowledge to support and further refine his explicit selection principle is that it was conveyed through face-to-face interaction. This interaction however implied a two-way knowledge flow: Darwin's close friends and colleagues also both conveyed and absorbed implicit knowledge through their frequent interaction. Before this argument is further developed, it is instructive to consider the prolonged period over which Darwin struggled with the difficulties he thought must be solved before his selection theory could be communicated to a wider audience.

## 3. DARWIN'S DIFFICULTY

This section points out that the picture of Darwin as the isolated genius absorbing explicit scientific knowledge contained in the books and journals of science is misleading. Darwin throughout his career engaged in social interaction, including with fellow scientists, and the peak of his social activities was in the years after he returned from the *Beagle* voyage (Richards 1987). The significance of Darwin's social activity for this chapter is the argument that Darwin through his extensive social interaction with the scientific community acquired the preconceptions required to make progress. That is, Darwin participated in meetings in small face-to-face groups that mostly included professional and amateur scientists. A consideration of the prolonged period during which Darwin struggled with the difficulties he thought must be solved before selection theory could be communicated to a wider audience thus illustrates how Darwin's personal contacts and social life were a source of inspiration. This further establishes the basis for a neo-Darwinian science model whose inheritance track is constituted by implicit and explicit knowledge operating in two distinct but connected trails. Not only was explicit knowledge conveyed among group members, but the recurring and intimate character of meetings in small face-to-face groups also favoured the mutual transfer of implicit knowledge. As discussed below, one of Darwin's difficulties relates to the ontological conflation of habits and instincts.[10] For us this conflates a genetic and a phenotypic trail of evolution. In consequence, this chapter outlines a model that aims at a careful distinction between the trails that constitute the process of inheritance in conceptual evolution.

Contemporary discussions often emphasise the inappropriateness of biological models for the explanation of rational economic behaviour (for example, Rosenberg 1994). This viewpoint is based on the idea that an enormous gulf divides creative human intentionality and dull instinct-following animal behaviour, as well as the observation that the life of economic actors involves a degree of complexity many orders of magnitude greater than cultural or biological life (see for instance Nelson 1995). Both ideas can be traced through the evolution of evolutionary thought since the 18th century (for example, Richards 1987; Worster 1994). Whereas the complexity argument, perceived as a difficulty for evolutionary models of thought, is of more recent origin, the perceived divide between the foundations of human and animal behaviour posed the single most formidable difficulty for the wider acceptance of Darwin's theory (Richards 1987). In contrast to the predominant Victorian conception of a divide between savage beast and civilised man, Darwin's selection argument implied continuity between the expressions of these two forms of biological existence (referred to as the 'continuity claim' in the following).

Whereas the intellectual climate in the 19th century favoured acceptance of Darwin's selection theory as the explanation of evolution in the animal kingdom, Darwin experienced mounting pressure against the application of his ideas to the evolution of human rational and moral faculties. These more formidable objections to Darwin's ideas came from supporters and friends rather than opponents who could muster less convincing counterargument (Richards 1987).

According to Richards (1987), the objections against Darwin's selection theory as explanation also of morality were based on the following three arguments. Lyell had observed that nothing like morality could be found in the animal kingdom, a fact suggesting that the chasm separating the animal and the human mind was simply too wide. William Rathbone Greg (1809–1881) and Francis Galton (1822–1911) moreover claimed that the evolution of a moral faculty would tend to benefit the unfit, a point elaborated by Wallace who argued that morality and intellectual capacity often resulted in disadvantage to survival because altruistic behaviour tended to harm those persisting in it. Lyell's observation, it is important to note, reflects the general objection to the 'continuity claim' in the 19th century. Whereas the continuity claim today turns on the question of rationality – 'Can animals reason as we do?' – the more significant question in 19th century England was: 'Can animals make moral judgements as we do? Or more pointedly: Is man essentially no more moral than a rutting pig?' (Richards 1987, p. 109). Although the rationality question played an important role for Darwin, the objections based on the morality question gave the impetus that drove Darwin to spend considerable effort in producing a convincing explanation for the evolution of morality (ibid.).

Darwin became familiar with Lyell's objection to the continuity claim aboard the *Beagle*, but had long before that time acquired an interest in moral philosophy. As noted above, Darwin absorbed Paley's work on moral philosophy at Cambridge, but as early as 1827, the year when he came down from Edinburgh, he had been introduced to moral philosophy. As often happened for Darwin, personal contact inspired literary contact:

> ... his introduction to Macintosh, as with Martineau, came through personal rather than literary contact. Macintosh was the brother-in-law of Darwin's uncle, Josiah Wedgwood, and like the young nephew he frequently visited the Wedgwood country house at Maer. Darwin first met him in 1827 and fell entranced under the power of the older man's intellect. In his *Autobiography* [p. 66], Darwin recalled that at the time he 'listened with much interest to everything he [Macintosh] said, for I was as ignorant as a pig about the subject of history, politicks and moral philosophy'. (Richards 1987, p. 115)

Harriet Martineau, mentioned in the above quote, was an essayist and radical social critic connected to the Wedgwood family through friendship with Fanny Wedgwood, the daughter of James Macintosh and wife of Hensleigh Wedgwood (Richards 1987). Darwin however first met Martineau in 1838 at a dinner party given by his brother Erasmus (ibid.). Richards (1987, p. 112) gives the following description of their meeting:

> Her intellectual charms bedazzled the young naturalist, but no less the other eminent minds of her widening circle, which included the irascible historian Thomas Carlyle, the great amateur scientist and Chancellor of the Exchequer Henry Loyd Brougham, and Britain's leading geologist Charles Lyell.

As the above examples indicate, the picture of Darwin absorbing explicit scientific knowledge contained in the books and journals of science is misleading. Darwin participated in meetings in small face-to-face groups that mostly included professional and amateur scientists, and through this social activity implicit as well as explicit knowledge was conveyed among group members. Already in Edinburgh, Darwin had befriended the physician and part-time lecturer Robert Grant, who discussed Lamarck with Darwin and inspired him to investigate marine invertebrates (Richards 1987). Although Darwin was supposed to study medicine, personal contacts with the community of natural scientists at the University of Edinburgh led him further astray towards the pursuit of an altogether different interest. Darwin became a member of the Plinian Society, where he presented a discovery he had made in marine biology (Richards 1987). Also in Edinburgh, he attended the Wernerian Society (after the great German geologist Abraham Werner), where the papers he heard further inspired him to enrol in the geology and zoology classes taught by the Society's founder, Robert Jameson (Richards 1987). About a decade later, during the summer and early autumn of 1838, the year of the Malthusian revelation, Darwin was also busy tracing out an evolutionary theory of morality (Richards 1987). According to Richards (1987), four books in particular helped Darwin along the way. Among these, there were two authored by people (Martineau and Macintosh) he had previously met through his social activities. Again, the implication is that Darwin's absorption of explicit codified knowledge through literary contact benefited from the implicit knowledge he had already absorbed through personal contact.

Although I have only cited a few of the examples Richards (1987) provides, I think they illustrate the point well, that Darwin's intellectual development was nurtured through personal contact with a tightly knit network of influential and closely connected colleagues and friends.[11] It was these contacts that allowed Darwin to refine and develop the conceptual categories he would need to fully comprehend and communicate his selection theory. But the personal contacts worked both ways: it was also Darwin's friends who posed the most formidable difficulty for Darwin's theory. Only Darwin's friends had the intimate knowledge required to point to the weak points of his theory.

So far, so good: to be persuasive to a wider audience, Darwin needed to counter the difficulty of moral continuity, but to persuade his friends and colleagues, he needed to overcome further obstacles. These obstacles can all be associated with the theoretical problems that, during Darwin's time, followed the absence of any empirical evidence on the medium and process of inheritance. According to the then prevailing inheritance theory, biological entities could develop *habits* by repetition of particular patterns of behaviour. These habits were somehow, over a number of generations, thought to develop into *instincts*. This conceptual apparatus involving habits and instincts was widely used and commonly accepted in the 19th century. Since it was further commonly acknowledged that the instincts were provided by direct interposition of the Creator, the challenge for Darwin, and Lamarck before him, was to provide a naturalistic explanation that excused the Creator in this matter. One difference between Darwin and Lamarck however was that Darwin had collected the evidence necessary to support his theory convincingly (see for example Mayr 1982). Another difference was that Darwin, for a number of reasons, delayed communication of his selection theory to a wider audience. Amongst those reasons, one of the most important was that he needed to solve the difficulty associated with instinct inheritance in the face of the widely held belief that the Creator implanted instincts at birth.

Only gradually and with much hesitation did Darwin arrive at a version of his selection theory that incorporated instinct inheritance (Richards 1987). To appreciate the severity of this difficulty one must remember that there were no genes and no neo-Darwinians around. The role played by instincts in the propagation of traits across generations of biological entities was simply a conjecture. Only with hindsight can we appreciate that the properties Darwin imputed to instincts were to some extent analogous to those we associate with genes. Darwin came close however in thinking of instincts as repositories of cumulated behavioural wisdom that tend to be preserved across generations according to the benefit they bestow on their carrier in the struggle for life. The problem however was that the mechanism of inherited habit that Darwin had considered early on remained active along with the principle of natural selection (Richards 1987). The unobservable nature of instincts can, therefore, be seen as the source of what Richards (1987) views as the three chief impediments responsible for Darwin's slow extension of his selection theory:

> The first was the ease with which habit could explain the origin of instinctive behavior. Not only did Darwin find this true, but so did Lamarck, Frédéric Cuvier, and John Sebright, all of whom he carefully read. The second difficulty, I believe, related to Darwin's lack of a clear distinction between animals selecting habits because of their usefulness in nature and nature's selecting animals because of their useful habits. ... From our perspective this confuses what we have come to think of as Lamarckian mechanisms with neo-Darwinian. ... [The third difficulty] was the near fatal problem that Kirby and Spence brought to his attention in their *Introduction to Entomology*. (Richards 1987, p. 143)

Since Darwin had no possibility of observing the stuff that preserved beneficial variation, he continued to view instincts as inherited habit, a possibility that could exist along with natural selection of instincts.[12] The first difficulty then follows from the lack of constraint on the post-hoc rationalisations that Darwin used to test his selection theory against observable evidence (for instance in *Origin*, Darwin discusses observations that show how birds in the United States and England acquire an instinctive fear of man). On the one hand Darwin argued that habitual action could become inherited. On the other hand, there was the possibility of natural selection of instinct:

> Under changed conditions of life, it is at least possible that slight modifications of instinct might be profitable to a species; and if it can be shown that instincts do vary ever so little, then I can see no difficulty in natural selection preserving and continually accumulating variations of instinct to any extent that may be profitable. (Darwin [1859] 1985, p. 236)

The second difficulty relates to the ontological conflation of habits and instincts. To repeat, there was no way Darwin could establish the required distinction between a genetic and a phenotypic track of evolution. As Richards (1987, p. 143) notes: 'Needless to say, Darwin was not a neo-Darwinian.' Had Darwin been a neo-Darwinian, he would have insisted that instincts and habits belong to two distinct ontological levels, one (instinct as equivalent to genes) constituting the other (habit equivalent to behavioural trait). Due to the impossibility of directly observing the process of inheritance, arguably the greatest conceptual difficulty for Darwin was that he had early on absorbed the idea, common at his time, that inherited habit could somehow become instinct. The crucial problem with this idea was the ontological conflation of habit and instinct, implying a conflation of what we would today refer to as genes and traits. The following excerpt from *Origin* provides a wonderful illustration:

> If we suppose any habitual action to become inherited – and I think it can be shown that this does sometimes happen – then the resemblance between what originally was habit and an instinct becomes so close as not to be distinguished. If Mozart instead of playing the pianoforte at three years old with wonderfully little practice, had played a tune with no practice at all, he might truly be said to have done so instinctively. (Darwin [1859] 1985, p. 235)

As argued below, previous models of scientific evolution have to some extent suffered from a difficulty similar to the one Darwin had encountered. Thus, most evolutionary models of science tend to conflate the notional genetic level with the notional level of traits (note here that the distinction between Darwinism and Lamarckism concerns a different issue). As explained in detail in the ensuing section, the present chapter aims to remedy this problem by application of neo-Darwinian wisdom. Before doing this however, we must briefly consider the third and most formidable of the difficulties associated with Darwin's selection theory of instinct. Darwin's third difficulty concerned the neuters or sterile females in insect communities, which he considered 'insuperable, and actually fatal to my whole theory'.

> ... for these neuters often differ widely in instinct and in structure from both the males and fertile females, and yet, from being sterile, they cannot propagate their kind. ... This difficulty, though appearing insuperable, is lessened, or, as I believe, disappears, when it is remembered that selection may be applied to the family, as well as to the individual, and may thus gain the desired end. ... Thus I believe it has been with social insects: a slight modification of structure, or instinct, correlated with the sterile condition of certain members of the community, has been advantageous to the community: consequently the fertile males and females of the same community flourished, and transmitted to their fertile offspring a tendency to produce sterile members having the same modification. (Darwin [1859] 1985, pp. 258–9)

Richards (1987) provides a detailed description of Darwin's struggle towards providing a solution to the problem of sterile insects. Being sterile precluded the possibility that such insects could pass on any favoured habits or instincts. At first the whole problem seemed to support the alternative to Darwin's theory of natural selection: the sterile condition was compelling evidence in favour of direct intervention by the Creator. Richards (1987) shows that Darwin had no satisfactory solution to the problem of neuter insects in 1848, and as late as 1857 considered several different explanations. Interestingly, Darwin as early as 1848 considered community selection, but rejected this solution on grounds of implausibility (Richards 1987). Why did Darwin reject community selection in 1848? Richards quotes a handwritten manuscript by Darwin from 1848 which indicates that:

> ... he [Darwin] had up to this time retained his older mechanism of hereditary habit as an integral element of his concept of instinct. (Richards 1987, p. 145)

To arrive at community selection would require Darwin to overcome the difficulty of conceptualising instinct inheritance without the interference of habit (since sterile insects could not pass on their acquired habits). Although Darwin considered community selection in 1848, part of the reason why he had to reject the idea then was that his established concepts of inheritance rendered this solution impossible. I believe this example shows that an important reason for Darwin's delay was that he had to overcome the difficulty of changing what had early on, by social interaction and reading, become his implicit assumptions regarding the nature of inheritance. In Chapter Seven of *Origin*, I think the trace of Darwin's difficulty in changing the conceptions of instinct inheritance he had already established is clearly present. The chapter starts out by outlining and endorsing the theory of inherited habit (which implies a conflation of habit and instinct) and ends with Darwin recognising that his solution to the problem of neuter-insects implies an alternative inheritance mechanism consistent with natural selection:

> The case [of neuter insects], also, is very interesting, as it proves that with animals, as with plants, any amount of modification in structure can be effected by the accumulation of numerous, slight, and as we must call them accidental, variations, which are in any manner profitable, without exercise or habit having come into play. For no amount of exercise, or habit, or volition, in the utterly sterile members of a community could possibly have affected the structure or instincts of the fertile members, which alone leave descendants. I am surprised that no one has advanced this demonstrative case of neuter insects, against the well-known doctrine of Lamarck. (Darwin [1859] 1985, pp. 258–9)

Here at last, Darwin had arrived at a conception of instinct inheritance consistent with natural selection in which the insect community was the unit of selection. Not only did his new conception of inheritance challenge his own previous conceptions, it was a case that clearly refuted the commonly-held belief in the 19th century, that inheritance was Lamarckian; in other words, that characteristics acquired by use and disuse were passed on to offspring. I believe this example shows that an important reason for Darwin's delay was that he had to overcome an enormous difficulty in changing what had early on, by social interaction and reading, become his implicit assumptions regarding the nature of inheritance. The point I hope to have made is that a crucial reason for Darwin's delay in publication of *Origin* is that he needed time to overcome the difficulty of neuter insects. And in order to do this he had to change the conception of inheritance he had acquired early in his career. That is, Darwin's explicit selection principle devised in 1838 required further development, in terms of rejecting old assumptions as well as refining the emerging conceptions which were consistent with natural selection. And in order to accomplish this, Darwin had to change both explicit and implicit conceptions. As further implied in the above, Darwin's social activities somehow helped develop the required implicit knowledge that allowed him to change previously held conceptions regarding the role of instinct. Consistent with the above picture, Hodgson (1993) concludes in his review of the multiple influences on Darwin, that it was the combined influence of (possibly implicit) social circumstances and explicit knowledge in written material that formed Darwin's thinking.[13]

In the ensuing section, I outline a neo-Darwinian model of science whose inheritance track is constituted by implicit and explicit knowledge operating in two distinct but connected trails. This model is consistent with previous evolutionary models and also shares the tendency in most evolutionary models to balance the weight of internal and external components in explaining scientific evolution. The chief difference is that the present model views implicit knowledge acquired by means of social interaction as a precondition for the expression of explicit scientific knowledge and outlines a neo-Darwinian inheritance track that employs both forms of knowledge.

## 4. EVOLUTIONARY MODELS OF SCIENCE

This section introduces the proposed neo-Darwinian science model. To do this, some terminology must first be established. I start out by defining 'replicator' and 'interactor' as general abstract analogues of 'gene' and 'organism'. Then I proceed to identify the explanatory features that must be shared by any evolutionary model and add three criteria that make such a model neo-Darwinian or neo-Lamarckian. Finally, a brief definition of what should be understood by 'evolution' and 'selection' is provided. Having thus cleared the way, I briefly review four of the most prominent evolutionary models of science (Hull 1990; Popper 1979; Richards 1987; Toulmin 1972). Against this background, the proposed neo-Darwinian model of science is presented as a possible development.

To facilitate generalisation, I suggest that Hull's (1990) definitions of replicators and interactors are adopted as abstract analogues of genes and organisms.[14] A *replicator* is 'an entity that passes on its structure largely intact in successive replications' (Hull 1990, p. 408). According to Dawkins (1976), good replicators are characterised by *longevity* (potential immortality through copies even if the individual copy has short life), *fecundity* (a high number of copies) and *fidelity* (accurate production of copies). An *interactor* can be defined as: 'an entity that directly interacts as a cohesive whole with its environment in such a way that this interaction *causes* replication to be differential' (Hull 1990, p. 408). Usually, the evolutionary process is explained in terms of the tripartite mechanisms of variation, selection and retention (Campbell [1974] 1987). In view of the need to develop an adequate description of the entities involved in the evolution of science, the term replicator and interactor are particularly attractive as supplementary characteristics since they help to carefully consider the nature of the entities that evolve.[15]

Evolutionary models of science as a knowledge-accumulation process share three explanatory essentials (variation, selection and retention) with any model of an evolving system. They must have replicating units that preserve and/or propagate selected variations; consistent selective retention of some variants at the expense of others; and sustained recreation of variation at part of the replicating units (Campbell [1974] 1987; Durham 1991). The assumption is that there must be an analogy between *any* models of evolutionary change, including evolutionary models of science, at some level. This viewpoint is summarised in what Durham (1991) has termed Campbell's Rule: the analogy between cultural accumulations is not in biology *per se* but rather in the general model of evolution of which organic evolution is but one instance (Campbell 1969). As Campbell ([1974] 1987, p. 56) further notes: 'In general the preservation and generation mechanisms are inherently at odds, and each must be compromised.' This is within the context of organisational learning that March (1991) refers to as balancing exploitation and exploration.

To make this qualify as a neo-Darwinian model, I further propose the following criteria. First, the model in question must have an inheritance track that includes two distinct trails defined by replicators and interactors such that information can only flow from replicators to interactors, never the reverse.[16] Second, the distribution of interactors should map one-to-one onto the distribution of replicators. Finally, replicators are temporally prior to, and constitute, interactors. A neo-Lamarckian model is characterised according to similar criteria with the exception that interactors can adapt their replicating code and this adaptation is conveyed to new generations of interactors. Due to problems in the neo-Lamarckian model associated with securing the basis for consistent selection, referred to as the baseline problem in previous work, the neo-Darwinian model must, as Campbell ([1974] 1987) insists, be fundamental to all genuine increases in knowledge – to all increases of system to environment.[17] Finally, to overcome the difficulty of explaining the supposedly 'Lamarckian' nature of much evolution in the social realm, I propose a solution, developed in previous work (Knudsen 2001), concurring with Campbell's outline of a nested hierarchy of selective retention processes.[18] As explained in more detail below, the proposed solution, referred to as Local Emulative Selection, adds a possible fundamental neo-Darwinian basis for the understanding of evolution in the social realm that somewhat extends Campbell's ([1974] 1987) work.

In biology, genes contain a code, which is replicated in a high-fidelity copying process. The new genetic code provides the instructions, which, dependent upon environmental triggers, unfold into the traits of the mature organism. The mature organism, in turn, comes with a potential to interact with the environment as a cohesive whole in a way that causes replication to be differential. In biology, this layered theoretical structure corresponds to empirical reality, the replicating code to genes, and the interacting entity to the organism. It is further important to emphasise that the relation between the code and its carrier is not deterministic. The code contains a wide range of potentials, which are gradually triggered by environmental cues through the organism's unfolding in maturation. In this sense, it works very much like a recipe for a mince pie. Although all mince pies have shared features they are also unique and none are quite like my grandmother's. In short, any organism is programmed by the genetic code; however, it is programmed to learn (Jacob 1985).

In biology, the code is contained in the genes in terms of a specific sequence of nucleotide bases, which make up the DNA molecule. In the social realm, matters are less settled but something like a 'cultural code' (Schultz 1994), social 'instinct' or 'habit' (Hodgson 1997), 'meme' (Dawkins 1976; Durham 1991), 'routine' (Nelson and Winter 1982), or 'replicator' (Hull 1990; Hull et al. 2001) may be used as analogy. The shared overlap of models employing these concepts is that an underlying code acts as recipe for the development of an entity (organism or agent) and its capacities. Evolution, then, can be explained in terms of two general population level subprocesses: (1) changes in the distribution of codes associated with replication; and (2) changes in the distribution of codes due to interaction. In biology, replication involves recombination and mutation (genetic response), whereas interaction involves imitation, competition, migration, differential reproductive success and death (phenotypic selection).

Following this scheme, scientific evolution by natural selection may be defined as a two-step process. Step one involves (direct) *replication* of a notional code (containing implicit and explicit scientific knowledge), and in step two the entity of interest (directly) *interacts* with the environment in a way that causes differential replication (of implicit and explicit scientific knowledge). This leads to a definition of the evolution of scientific knowledge in terms of variation accumulated over time because of the independent but causally linked subprocesses of knowledge replication and interaction among scientists. Selection, in turn, can be defined as 'a process in which the differential extinction and proliferation of interactors cause the differential perpetuation of the relevant replicators'. (Hull 1990, p. 409). Binmore (1990, p. 17) provides a good description of this process:

> In social evolution it is ideas which are the basic unit of replication but, instead of being replicated from one body to another by biological reproduction like genes, they are replicated from one head to another by the social processes of communication and imitation. Here an 'idea' may be a complex theory, or it might be a rule of thumb for achieving some purpose, or it might be little more than a learned reflex for action.

According to this view, scientists are knowledge-carriers, and according to their success in the scientific community (as measured for example in terms of professional status), their ideas will be differentially propagated. That is, selection operates on the population of ideas carried by scientists, and on their articles, books, research notes, databases and the like.[19] Note here that this model views the replication of explicit knowledge as a process distinct from, and yet nested within, the process of replicating implicit knowledge in face-to-face groups.

Having provided the necessary general terminology and understanding of evolutionary explanations, we can now continue with a brief review of the class of evolutionary theories of mind and behaviour to which evolutionary science models belong. Richards's (1987) description of the emergence of evolutionary theories of mind and behaviour offers a rich historical account that aims to follow the Darwinian tradition of balancing the extreme internalist (such as Collingwood 1956) and externalist doctrines (Bloor 1976) in the history of science. Although most historians of science blend internal and external components in arguing the course of scientific thought, there is a tendency to lean more or less towards either pole. Internalists trace the evolution of ideas, their internal logic and conceptual linkages as well as the shifts in evidential support, but tend to downplay social and psychological causes (Richards 1987). Whereas internalists tend to trace explicit scientific knowledge that lives in the disembodied realm of thought, externalists tend to embed scientific ideas in the human world (ibid.). That is, externalists trace the implicit psychological complexes and social relations that mediate or perhaps even determine the expression of scientific ideas.

According to the central thesis of this chapter, the internalists and the externalists each focus on one of two distinct expressions of scientific knowledge, implicit and explicit knowledge, operating in two (ontologically) distinct but connected trails. This thesis is the basis for the neo-Darwinian model of scientific evolution outlined in this chapter. As pointed out below, Toulmin (1972) and Richards (1987) offer related models that build on similar theses. By emphasising the evolutionary significance of implicit knowledge, this chapter however goes one step further towards the formulation of a neo-Darwinian model of scientific evolution in which the replicating code is the implicit knowledge that constitutes the scientist's conceptual categories. Although some models of science consider implicit knowledge (such as Polanyi 1966), for example in terms of the Kantian *a priori* categories that enable conception (Richards 1987), very few have considered implicit knowledge as a candidate for the replicating equivalent of DNA. One of the very few is Schultz (1994), and as Schultz points out, this idea can be found in Thomas Kuhn, Harold Bloom and Jacques Derrida.[20] In general this leads Schultz to conclude:

> Future philosophers of science, literary critics, and scholars in any field can benefit by using the idea of a cultural genetic code, for tradition forms do act much like DNA. ... Human culture has evolved beyond biological species, for it has a freedom of self-determination unknown in the merely biological sphere. Cultural genetic codes are thus more than biological genetic codes. The evolving of cultural genetic codes raised humans out of the merely biological order to a being of tradition. (Schultz 1994, p. 489)

Table 5.1, at the end of this chapter, provides a schematic representation of core features of the most influential evolutionary models of science. The shared feature of these models is their focus on the growth of scientific knowledge within a natural-selection epistemology. That is, they can be viewed as a subset of the large evolutionary epistemology programme founded by Popper (Campbell [1974] 1987). Table 5.1 therefore also includes Popper's (1979) influential model. As can be seen from the table, Popper (1979) to some extent black-boxed external elements, and focused primarily on the evolution of explicit knowledge (hypotheses) by means of rational criticism. Although perfectly consistent with a model of evolution by natural selection, it may be seen as a weakness that Popper is not very forthcoming in defining a mechanism of knowledge transmission (inheritance). I believe this assessment is wrong, however, since Popper's is a model of explicit knowledge which is obviously available in books, journals and so on. More reasonably Popper's model may be seen as an attempt to overcome the problem of the *a priori* (see, for example, Campbell [1974] 1987), that is, to provide a model that can ultimately explain the foundation of (natural) science.

What all models in Table 5.1 share is their conception of science as a selective system that adapts its explanation to the environment by a deliberate trial-and-error process. According to Popper (1979) the growth of scientific knowledge proceeds by natural selection of hypotheses. Explicit scientific knowledge is contained in the hypotheses held by individual scientists, and depending on their success in explanation of a problem, hypotheses will replicate differentially among scientists. That is, hypotheses serve as replicators, the general analogue of the gene. Popper's key assumption is that we start from problems. But where do problems come from? According to Popper we come with inborn knowledge, that is, inborn expectations or anticipations about the state of the world. But since this knowledge, according to Popper, is by definition inadequate, our first problems come from the disappointment of inborn expectations. Having explained how the first problems emerge, Popper (1979, p. 259) then describes the ensuing growth of knowledge as 'consisting throughout of corrections and modifications of previous knowledge'. And it is in the course of modifying hypotheses, which contain our expectations in the form of explicit conjectures, that variation is produced. Scientists produce inadequate solutions (conjectures) to scientific problems, adopt mistaken beliefs, and are led to explore new problems and solutions by curiosity. The growth of knowledge, therefore, 'proceeds from old problems to new problems, by means of conjectures and refutations'. (Popper 1979, p. 258). The final element of Popper's model is the criterion of selection. The hypothesis will survive if it can withstand rational criticism and is not refuted by empirical evidence.

As can be seen from Table 5.1, Toulmin (1972), Richards (1987) and Hull (1990) agree in defining scientific ideas as replicators and the scientists as the interactors, the carriers of ideas. What these models add to Popper's replicators (hypotheses) is some of the stuff externalists usually emphasise – beliefs about the ultimate goals, general aims and explanatory ideals of science, proper ways to go about realising those goals, modes of representation, and so forth. Although blending internal and external elements of science, it must be emphasised, that these models all share Campbell's ([1974] 1987) description of science as a selective system:

> The demarcation of science from other speculations is that the knowledge claims be testable, and that there be available mechanisms for testing or selecting which are more than social. In theology and the humanities there is certainly differential propagation among advocated beliefs, and there result sustained developmental trends, if only at the level of fads and fashions. What is characteristic of science is that the selective system which weeds out among the variety of conjectures involves deliberate contact with the environment through experiment and quantified prediction, designed so that outcomes quite independent of the preferences of the investigator are possible. (Campbell [1974] 1987, p. 71)

That is, all the models in Table 5.1 characterise science as a deliberate activity which, in principle, can be designed so the preferences of the scientist do not influence outcomes of experiments and predictions. In sum, the four models all view science as a deliberate knowledge-accumulation process and provide the required account of the three explanatory essentials (variation, selection and retention). The models all define replicating units that presumably preserve and/or propagate selected variations. They account for presumably consistent selective retention of some variants at the expense of others, and they provide reasonable and detailed description of mechanisms that recreate sustained variation on the part of the replicating units. For two reasons, however, I propose to extend these models by including an additional inheritance mechanism. The first reason is that this allows the definition of a nested model whose foundation is consistent with the above stated requirement that it be neo-Darwinian. The second reason is, as illustrated in the above account of Darwin's delay and difficulty, that social interaction and possibly implicit knowledge acquired in face-to-face groups plays a crucial role in forming the scientist's presumptions and conceptual categories. That is, as illustrated by Hull (1990), Richards (1987), and many other historiographies, the scientist's presumptions, conceptual categories and psychological dispositions will both enable and constrain his or her deliberate efforts. This is no less true in the professions. As shown by the research presented in Sternberg and Horvath (1999), implicit knowledge is an integral part of law, military command, medicine, management, sales and teaching. As further implied in the above, Darwin's social activities somehow helped develop the required implicit knowledge that allowed him to benefit from reading and experimentation. Admittedly, this is a conjecture. As described in more detail below, it is a conjecture however that is consistent with experimental psychology (Reber 1993), psychology of the professions (Sternberg and Horvath 1999) and the psychology of human error (Reason 1990).

## 5. A NEO-DARWINIAN MODEL OF SCIENCE BASED ON LOCAL EMULATIVE SELECTION

The present section outlines a possible neo-Darwinian model of science. It is a nested model based on what I have previously (Knudsen 2001) referred to as local emulative selection. This principle adds an additional foundational layer to the above models of science. This layer consists of replicators that contain implicit knowledge outside the reach of consciousness. Within this layer, there is nested a second level of replicators containing explicit knowledge and a third layer, which contains the *professional identity* of the interacting scientist.[21] That is, the scientist's professional identity is in this model the interactor, and this professional identity is constituted by both implicit and explicit knowledge. To establish the argument, I need to distil two key insights from recent empirical research on implicit knowledge. Armed with this support I then proceed to outline the proposed model.

During the last twenty years, the idea of implicit learning has gained influence due to massive empirical support from rapidly growing research in experimental psychology (Kentridge and Heywood 2000; Liebermann 2000; Nisbett and Wilson 1977; Reber 1993). Using increasingly sophisticated experimental design, it has consistently been shown that cognitive schema develop without awareness in a process referred to as implicit learning (Kentridge and Heywood 2000; Reber 1993). Implicit learning involves long-term exposure to stimuli, does not necessarily involve meta-cognition and must be distinguished from transient activation (Bock and Griffin 2000). It is also important to note that implicit learning can involve both improvement and detrimental outcomes (Woltz et al. 2000). Based on a number of empirical studies in the workplace, Eraut (2000) divides the tacit knowledge conveyed through implicit learning into three categories: (1) tacit understanding of people and situations; (2) routinised actions; and (3) tacit rules that underpin intuitive decision-making. According to Eraut (2000), these three types of tacit knowledge come together when professional performance involves sequences of routinised action punctuated by rapid intuitive decisions based on a tacit understanding of the situation. This leads to an understanding of intuitive, analytic and deliberative cognition as distinct and yet interdependent modes of cognition (ibid.), that is, as summarised in the support-hypothesis, analytic and deliberate cognition is supported by intuitive decision-making based on tacit knowledge conveyed through implicit learning.

Two insights central to the present argument can be distilled from this literature. The first concerns the nature of implicit knowledge: analytic and deliberate cognition is supported by intuitive decision-making based on implicit knowledge. That is, consistent with the support-hypothesis, explicit codified knowledge must always be supported by uncodified implicit knowledge. It has further been shown that cognitive schema to some extent develop without awareness. That is, Polanyi's (1966) claim that the scientist's conceptual categories also partly consist of implicit or tacit knowledge is supported by recent experimental psychology.

As argued above, an important reason for Darwin's delay may be that he had to overcome an enormous difficulty in changing what had early on, by social interaction and reading, become his implicit assumptions regarding the nature of inheritance. And in order to overcome the difficulty of neuter insects, he had to change those partly implicit conceptions acquired early in his career. To further develop the selection principle that presented itself in explicit form in 1838, Darwin needed to reject old assumptions and refine his emerging conceptions. Further experimentation and reading were certainly important for Darwin, but as suggested above, an additional reason for the long delay may have been that he needed to develop the implicit knowledge required to support his selection principle. At least this conjecture is consistent with Darwin's extensive and enduring social interaction with a number of the leading scientists of his day. This conjecture is however also consistent with the second insight of the above-mentioned studies, concerning how implicit knowledge is transferred. According to this insight, the transfer of implicit knowledge among people (presumably including scientists) involves long-term and direct exposure to stimuli; in other words implicit knowledge is directly transferred among members of face-to-face groups through long-term interaction.

Armed with this justification, I can now outline the proposed model. The key assumption of this model is the support-hypothesis: explicit codified knowledge is always supported by uncodified implicit knowledge. The so-called uncodified implicit knowledge constitutes the conceptual categories without which the explicit codified knowledge will remain inaccessible and impossible to communicate. The implicit knowledge and the explicit representations of science constitute two distinct, but mutually dependent and interacting, trails of inheritance.

Note also that according to the support-hypothesis it is wrong to see explicit knowledge as knowledge which, perhaps with some difficulty, has somehow been moved from an implicit to an explicit mental 'box'. Rather I believe that implicit knowledge always remains implicit even if we can observe seemingly successful instances of codification (see for example, Nonaka and Takeuchi 1995). This belief is based on the following idea: what we call codification (or decoding implicit knowledge) involves a change of kind and not merely the relocation of some knowledge from an implicit to an explicit 'box'. As explained in the following paragraphs, this means that codification will increase rather than reduce the amount of implicit knowledge.

The basic idea is that the role of implicit knowledge in codification is to act as a seed. And in the event of what may be viewed as a successful instance of codification, this seed will give rise to an explicit form of knowledge that may be reflected upon, communicated and further developed. At the very first instance of codification, the person(s) responsible for the production of some form of explicit knowledge (say a scientific hypothesis) may only conceive of an inkling of the full potential of this knowledge. The history of science is ripe with examples of the initial limited understanding of new explicit knowledge, such as Darwin's gradual realisation of the potential of his selection hypothesis from the moment in 1838 when it was conceived and throughout the rest of his life (see Richards 1987). According to the idea that implicit knowledge sometimes and somehow invokes the creation of novel explicit knowledge, it would be wrong to see explicit knowledge merely as decoded implicit knowledge. By contrast, the role of implicit knowledge in creative acts is better seen as a seed that gives rise to explicit knowledge. And should the explicit knowledge start growing, the underlying implicit knowledge, which is the basis for its further development and communication, must also continue to grow. If this is true, the successful codification of implicit knowledge will result in more and not less implicit knowledge.

The argument so far is that the evolution of knowledge in explicit form is constituted by implicit knowledge acting as support as well as catalyst. Without implicit knowledge, the formulation and communication of explicit knowledge are simply impossible. Furthering the development of explicit knowledge, in turn, will induce a growth in the underlying implicit knowledge that forms the categories necessary for its conception. According to this argument, understanding the evolution of explicit knowledge requires an understanding of implicit knowledge. As shown in schematic form below, what I propose is a model in which the role of implicit and explicit knowledge in the process of knowledge accumulation is to constitute two (ontologically) distinct but connected trails that constitute a complete inheritance track.

The first step is to define selection of explicit knowledge. I will use the standard definition used in the above models and define selection of explicit scientific knowledge as the gradual and slow change in the distribution of scientific ideas caused by their differential social transmission.[22] This process may be termed Lamarckian (see Figure 5.1A, at the end of this chapter): the individual scientists may, deliberately or not, change explicit knowledge (hypotheses, conjectures, elements of theories and so forth). Given the assumption that scientific knowledge can be subdivided into ideas influencing their carrier's behavioural traits, evolution can happen due to selection of ideas or because of a rapid but large change in the population of ideas. In the latter case, there will be some evolution but it is likely to be an autocatalytic process, which dies out fast (see for example Maynard Smith and Szathmáry 1999). Therefore, the gradual but slow selection of explicit scientific knowledge, which is consistent with neo-Lamarckian selection as shown in Figure 5.1A, will have a larger evolutionary potential than the adjustment associated with a large and fast change in the distribution of scientific ideas. Finally, a measure for fitness can be defined on the basis of the relative transmission success in Durham's (1991) definition of cultural fitness. Thus, Durham (1991) defines cultural fitness as:

> ... an allomeme's expected relative rate of social transmission and use within a subpopulation, where the 'expected rate' can be defined, following Philip Kitcher (1985, p. 51), as 'the probabilistically weighted average of all possible values'.

In similar terms the fitness of a scientific idea is defined as a scientific idea's expected relative rate of social transmission and use within a subpopulation (a discipline or research team), where the expected rate is the probabilistically weighted average of all possible values. So far, I have used Durham (1991) to describe scientific evolution as a particular instance of cultural evolution.

There is one problem, however, with Durham's (1991) theory and any other Lamarckian theory. In previous work I referred to this as the baseline problem (Knudsen 2001; 2002). Due to the possibility that the carrier of explicit scientific knowledge can change it, its evolutionary potential may be severely compromised. As mentioned above this point concurs with Campbell's ([1974] 1987) insistence that a blind-selective-retention process is fundamental to all increases in knowledge.[23] That is, unless the carrier of explicit knowledge is somehow constrained, the selection environment may not be sufficiently consistent. A possible solution to this problem is, as suggested by Campbell ([1974] 1987) and as explained in previous work (Knudsen 2001), a nested selection-retention hierarchy with the most fundamental being strictly neo-Darwinian. The second reason to include implicit scientific knowledge is that this form of knowledge seems a ubiquitous aspect of all forms of knowledge cumulation. This point was supported by recent evidence from experimental psychology and illustrated in the above account of Darwin's delay and difficulty. Further motivation for the proposed local selection of implicit knowledge by emulation can be found in *Origin*. Here Darwin repeatedly describes the notion of unconscious selection. The proposed principle of local emulative selection described in the following may be viewed as a particular instance of what Darwin referred to as unconscious selection.

In the proposed alternative, local emulative selection, transmission is implicit in the sense that ideational units are unconsciously emulated during a period of exposure to those of the transmitters. Thus, I define local emulative selection as the gradual and slow change in the distribution of implicit scientific knowledge caused by differential emulation (implicit social transmission) of tacit knowledge in face-to-face groups. The term tacit knowledge covers a continuum from knowledge not reachable by consciousness to completely conscious but inexpressible knowledge. Local emulative selection refers to tacit knowledge that cannot be reached by consciousness and thus involves implicit transmission of unconscious tacit knowledge components. This process is clearly neo-Darwinian (see Figure

5.1B at the end of this chapter): the individual scientist is precluded from changing implicit knowledge (the perspective which structures cognitive categories and schema). By contrast, cultural selection involves explicit transmission of conscious knowledge components. In cultural selection, the transmission is explicit in the sense that ideational units are chosen, imposed; or perhaps transmission more commonly involves a mixture of choice and imposition (Durham 1991). Note further that implicit knowledge refers to an ideational component expressed in a certain type of behaviour. Implicit scientific knowledge can, therefore, be seen as a habit of thought, an individually held ideational component acquired through emulation in a social setting which instils a disposition to act as well as a sensitivity to what type of behaviour is appropriate in specific circumstances. Recently, Hodgson (1997) has derived a similar point based on the term 'habit' taken from instinct psychology and the American pragmatists.

A further conceptual clarification is needed in order to align the term 'habit of thought' with Nelson and Winter's (1982) 'routine'. One possibility is to define routines as the social-level expression of habits (of thought or action). I have previously argued that the relevant social level is the face-to-face group, for it is at this level that implicit knowledge is conveyed. This also seems to be what Nelson and Winter (1982) write about, in other words, routines acquired through socialisation serve as organisation-level targets for control, trust and memory, and so forth. A routine then, is an ideational component acquired in a social setting (face-to-face group), which is expressed as a disposition to act as well as a sensitivity to what type of behaviour is appropriate in specific circumstances. A set of routines containing productive knowledge (ideational components) is, through implicit learning in a particular face-to-face group, passed on to a newcomer entering a business organisation. As a result the newcomer will, over time, acquire a set of regular behaviour and thought patterns suitable for the task at hand.

In similar terms, implicit scientific knowledge can be defined as routines (at the level of face-to-face groups) and habits of thought (at the individual level) passed on to a newcomer entering a research institution (university) or scientific community through implicit learning in a particular face-to-face group (research team, informal group). This implicit scientific knowledge is expressed as a cognitive disposition, a disposition to act as well as sensitivity to what type of behaviour is appropriate in specific circumstances. That is, the implicit scientific knowledge supports explicit scientific knowledge by enabling: (1) a tacit understanding of new scientific material contained in books, journals and so on; (2) routinised semi-automatic actions; (3) tacit rules that underpin intuitive decision-making used in determining what is relevant; and (4) a tacit understanding of people and situations. Note that these enabling conditions also define constraints in terms of the implicit perspective or frame of mind held by the scientist. The scientist will only consider explicit knowledge within the reach of his or her implicit perspective, as illustrated by Darwin's consideration of inheritance within the habit-instinct perspective he had held so long. I am assuming that part of this perspective was held implicitly – if not, Darwin's long delay seems inexplicable as well as his discussion of instinct and habit in *Origin*.

Explicit scientific knowledge is conveyed, as it always was, through books, journals, professional meetings and the like. And scientists may obviously choose to change their explicit theories and conjectures. The model proposed here however implies that the evolution of explicit scientific knowledge proceeds along a trail nested within the evolution of implicit knowledge. As indicated above this means that new implicit knowledge, acquired in a face-to-face meeting, may inspire development of new explicit knowledge expressed at some later stage. Similarly, new explicit knowledge, for example encountered in a book, can only be appreciated if the underlying implicit knowledge is to some extent already in place – alternatively, new explicit knowledge may inspire further development of previously held implicit knowledge. The following concluding section elaborates on the further implications of the proposed model.

## 6. CONCLUSION

The present chapter used Darwin as an illustration to show that both explicit knowledge read in books as well as implicit knowledge conveyed in social interaction among a small group of prominent scientists helped the creation, development and communication of his selection theory. I examined Darwin's delay in publication of *Origin* as well as the difficulties that caused this delay. On this basis I argued that Darwin's delay was partly caused by his need for time to acquire the implicit knowledge necessary to further develop his insights and to convey his results to the scientific community. Although this latter statement is admittedly a conjecture, it is consistent with our knowledge about Darwin's otherwise inexplicable difficulty in changing the conceptions of inheritance that he had acquired early on. It is also consistent with an emerging body of evidence from modern experimental psychology.

The primary reason to include implicit scientific knowledge in the proposed model is that this form of knowledge seems a ubiquitous aspect of all forms of knowledge accumulation (for instance, dog-breeding in the 19th century, professionalisation in the 20th century and so on), including the evolution of scientific knowledge. In order to include the evolution of implicit knowledge as a distinct feature, I proposed a nested neo-Darwinian model of science based on local emulative selection (the unconscious selection-of-knowledge structure that takes place in social interaction by way of emulation, for example, of excellent scientists). In this model the inheritance track is constituted by implicit and explicit knowledge operating in two (ontologically) distinct but connected trails, and thus includes some but far from all of the items in Campbell's ([1974] 1987) nested hierarchy of selective-retention processes. An additional reason for the proposed model was that its strictly neo-Darwinian basis concurs with Campbell's ([1974] 1987) insistence that a blind selective-retention process is fundamental to all increases in knowledge. Finally, in *Origin*, Darwin himself describes the notion of unconscious selection in terms consistent with the proposed principle of local emulative selection. The process of local emulative selection proposed here, and in previous work (such as Knudsen 2001), involves the unconscious emulation of thought structure in face-to-face groups. This implies that evolution of scientific knowledge, for example, might be driven by the competition for position in research groups admired for their members' professional excellence.[24]

In the following I consider further implications of the proposed model as a development of previous evolutionary models of science. First however I must point out that the transmission of explicit and of implicit knowledge respectively differ in the proposed model. The explicit trail, in principle, works by continuous transmission of scientific knowledge, a feature which is consistent with cultural accumulations in general (see Durham 1991). The implicit trail however works by discrete transmission. Only when scientists meet in a particular face-to-face group is implicit knowledge conveyed. This condition has further implications for the evolution of implicit knowledge: high stability in the composition of face-to-face groups and a high frequency of their meetings favour evolution of implicit knowledge. Note also that since newcomers bring new implicit knowledge, they will be a crucial source for continued creation of variation in implicit knowledge. According to this implication of local emulative selection, the *lone* genius should be a myth (or post-hoc rationalisation as suggested by Campbell ([1974] 1987). Rather we should expect genius and extraordinary creativity to be associated with highly stable (with respect to its membership) and relatively small research communities whose members interact very frequently. According to the present model, think-tanks and the like are good for the added reason that they stimulate the evolution of implicit scientific knowledge and thereby continued evolution of explicit scientific knowledge (it also helps that the best talent tends to gather in think tanks).[25] In general the possibility of vicariously managing the evolution of scientific research through the management of research groups is unwarranted however in terms of securing particular outcomes (March 1999). Although the right conditions in the stability and composition of research groups can secure the possibility of continued evolution (which is quite an accomplishment), the indeterminism of evolutionary processes should prevent more radical engineering approaches to the control of evolution (March 1999).

A further implication of the present model is that local emulative selection will slow down the evolutionary process relative to its speed in the direct transmission of explicit knowledge. As Binmore (1990) notes, the transmission of ideas (packages of knowledge) from head to head simply by imitation would be enough to fuel the evolutionary process. This however ignores more rational aspects of social evolution. According to Binmore (1990, p. 17), some ideas ('meme packages') in human societies, in particular rational or scientific ideas, 'owe their success to the fact that they keep their carriers where it matters by providing them with a more powerful method of acquiring useful memes than simple imitation'. In turn, such powerful methods of acquiring useful scientific knowledge will tend to accelerate the speed of evolution considerably (Binmore 1990). In addition there is the fact that explicit knowledge can be rapidly transmitted through a multitude of channels (Durham 1991) including the enormous acceleration of transmission rates made possible by electronic mail and internet resources only 'one click away'. Since local emulative selection will considerably slow down evolution because it takes place in relatively small face-to-face groups over long periods of time, it is important to distinguish between new explicit knowledge supported by previously held implicit knowledge and new explicit knowledge that also requires development of new implicit knowledge.[26] In the latter case, associated with shifts in technological progress and significant scientific breakthroughs, the present model suggests it would be a mistake to think that the evolution of significant scientific breakthroughs can be much speeded up.

Finally, I think it is worth pointing out that the proposed model includes both internalist and externalist views on the history of science. That is, implicit knowledge captures social aspects of science usually belonging to the externalist pole whereas the form of explicit knowledge considered here leans toward the internalist pole. As I hope the brief illustration of the history of Darwin's struggle with selection theory has shown, both social and private processes of knowledge acquisition are essential for understanding the evolution of scientific knowledge. Such tendencies to balance internalist and externalist views also characterise most previous evolutionary models of science. Toulmin (1972), for example, consistent with the present model's view, focused on the importance of enculturation: the intellectual techniques, procedures, skills, and methods of representation that were conveyed through apprenticeship in a scientific discipline. But he also considered the transmission of the substantive contents of science. Also Richards (1987) and Hull (1990) strike a balance between internal and external aspects of science. The difference however between these models and the one proposed here is that the present one, for reasons given above, aims to include a presumably important aspect of enculturation, the absorption of implicit knowledge, as a distinct feature. Clearly we need to know much more about the nature of implicit learning, but I believe we know enough to support the conjecture of local emulative selection which is the basis for the neo-Darwinian model proposed here. Another question is whether adding too much complication to existing models handicaps the proposed model. This is partly a matter of taste. I believe however that the importance of implicit knowledge also in science requires that we develop evolutionary models that include this form of knowledge as an explicit feature. The present model should be seen as an attempt to proceed in this direction.

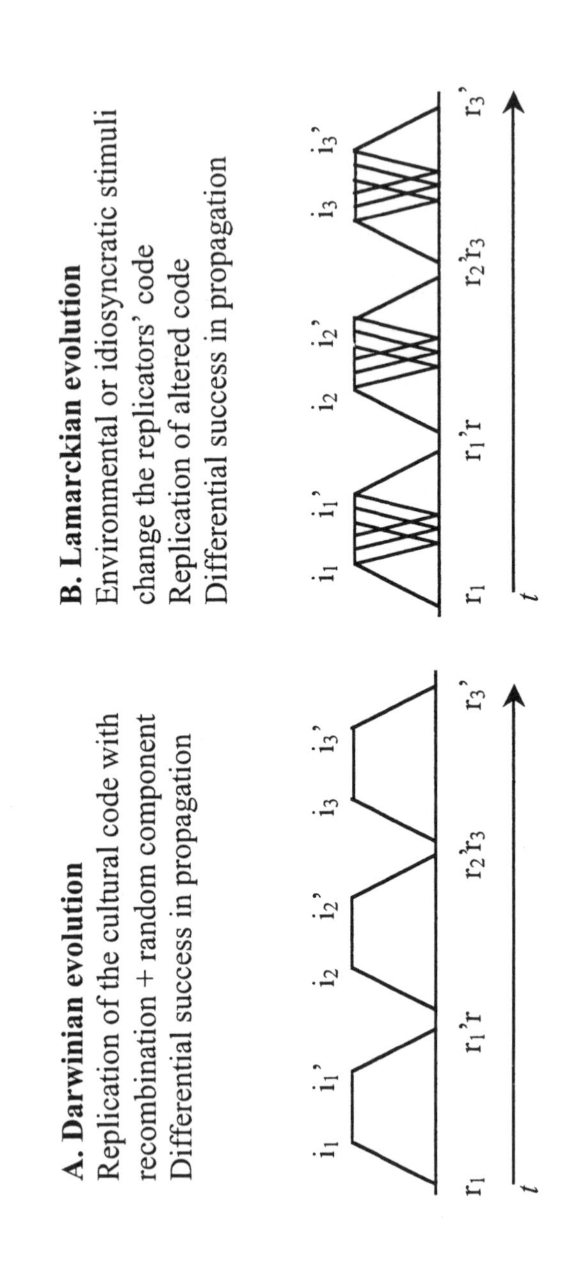


*Figure 5.1A Darwinian and Lamarckian evolution*

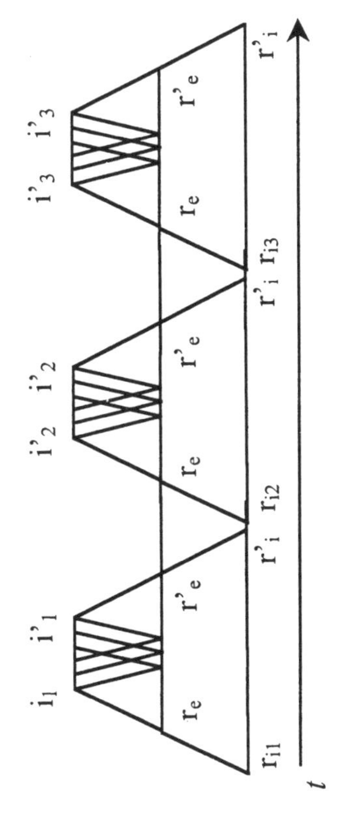


*Figure 5.1B Neo-Darwinian model of science based on local emulative selection*

*Table 5.1 Comparison of influential evolutionary models of science*

| Element | Toulmin (1972) | Popper (1979) | Richards (1987) | Hull (1990) | LES model |
|---|---|---|---|---|---|
| Replicators | Competing intellectual variants in terms of: (1) substantive content, (2) ultimate goals, general aims, explanatory ideals of science | Scientific hypotheses | Specific ideas of a theory united into genotypes by the bonds of logical compatibility and implications and ties of empirical relevance | Elements of the substantive contents of science: beliefs about norms, goals of science, modes of representation, accumulated data | Implicit and explicit knowledge entities regarding substantive content, method and aims of a particular scientific discipline |
| Interactors | Scientists (members of scientific disciplines) | Scientists | Scientists' cognitive representations of a scientific theory | Scientists' cognitive representations of a scientific theory | Professional identity of scientist |
| Variation | Innovative ideas emerging through rational problem-solving, institutional demands, social interests | Production of inadequate solutions (conjectures) to scientific problems, mistaken beliefs, curiosity, responses evoked by challenges from the environment | Cognitive representations of a scientific theory vary according to the slightly different ideas constituting it, their relations, and the changing intellectual and social environment that supports it | Recombination of ideas and invention, misunderstanding, differential intellectual interests, differential 'extra-scientific interests' including career interests | Recombination of explicit (intentional effort) and implicit knowledge components (changes in face-to-face group membership and/or characteristics), emergence of new explicit knowledge components |
| Transmission | Enculturation: intellectual techniques, procedures, skills, and methods of representation employed in explanation conveyed through apprenticeship in scientific discipline | Assumed in terms of the integration of knowledge over time | Replication of ideas from books and empirical observation | Replication of ideas from books, journals, computers, and human brains | Emulation of implicit scientific knowledge in terms of conceptual categories acquired face-to-face in group, replication of explicit scientific knowledge from books, journals etc. |

*Table 5.1 continued*

| Element | Toulmin (1972) | Popper (1979) | Richards (1987) | Hull (1990) | LES model |
|---|---|---|---|---|---|
| Transformation | Differential propagation of local changes in ultimate goals, general aims and explanatory ideals within a specific scientific discipline | Corrections and modifications of previous (in principle fallible) knowledge by means of conscious and systematic rational criticism | Differential transmission of ideas constrained by the vagaries of education and intellectual connections, the social milieu, psychological dispositions, previously settled theory, and recently selected ideas | Differential transmission of ideas, formation and expansion of research groups | Differential replication of implicit (changes in face-to-face group membership and/or characteristics) and explicit knowledge components (corrections and modifications of previous knowledge by means of conscious criticism/interests) |
| Isolation | Limited scope of scientific discipline | Pure versus applied science | Distinct conceptual systems (including theoretical concepts, methodological prescriptions or general aims | Scientific groups ('demes') | Face-to-face group, scientific community |
| Selection criteria | Scientific discipline, wider social interests | Scientific community: fallible conjectures (anticipations or expectations), rational criticism and refutation | Scientific norms (logical consistency, semantic coherence, standards of verifiability and falisifiability, and observational relevance), individual scientist, scientific communities, public scrutiny | Scientific community, social interests | Intra- and extra-scientific preferences of face-to-face group, scientific community, and social interest groups |

# NOTES

1 A draft that contained some of the present argument was presented at the conference 'Science as a Spontaneous Order', Skagen, 16–18 January 2000. Inspiration from participants at this event stimulated the development of the present chapter. The author is also very grateful for stimulating comments offered by Geoffrey M. Hodgson, David L. Hull and Robert J. Richards. All remaining error was produced without any help.
2 Since 'explicit knowledge' is reasonably viewed as an emergent property of 'implicit knowledge', I present 'implicit knowledge' and 'explicit knowledge' as *ontologically distinct*. This is not to say that 'implicit knowledge' and 'explicit knowledge' can be viewed as two extremes of a possible continuum rather than polar opposites.
3 By emulation I refer to a tendency to form knowledge structures that are somewhat similar to those of other people. As explained below, emulation will be more effective under frequent interaction in relatively stable face-to-face groups – and the unconscious part refers to what experimental psychologists (such as Reber 1993) call implicit learning, the demonstrable transfer of knowledge outside the reach of consciousness.
4 The purpose of this chapter is not to present new historiographic material. Rather, known material is used as an illustration for the proposed model of co-evolving implicit and explicit scientific knowledge. The historiographic material in the sections 'Darwin's Delay' and 'Darwin's Difficulty' is to a large extent based on Richards's (1987) fine historiography which in turn includes a very large number of sources. I also draw on Mayr (1982) and Darwin's *Origin*.
5 In the Introduction to *Origin* (First edition), Darwin writes: 'My work is now nearly finished; but as it will take me two or three more years to complete it, and as my health is far from strong, I have been urged to publish this Abstract. I have more especially been induced to do this, as Mr Wallace, who is now studying the natural history of the Malay Archipelago, has arrived at almost exactly the same conclusions that I have on the origin of species.' (Darwin [1859] 1985, p. 65).
6 See Turner (1994) for a useful overview of the extensive philosophy literature on this topic.
7 See Hodgson (1993) for a fine review of the history of the two-way flow of ideas between 'natural economy' and 'political economy' in the late 18th century as well as the 19th century. See also Schweber (1980) and the material in Kohn (1985) and Mirowski (1994).
8 At Cambridge, Darwin made friends with John Henslow, whose lectures on botany he enjoyed. Henslow introduced Darwin to William Whewell and Adam Sedgwick and secured for Darwin the position on HMS *Beagle* (Richards 1987).
9 Chambers was criticised for not proving his naturalistic theory of species transmutation with descent, which involved a claim of continuity between man and beast (Richards 1987).
10 Darwin never completely overcame this difficulty (see for example Darwin [1859] 1985, Ch. Seven), mainly for the reason that the medium and mechanism of inheritance could not be observed at his time.
11 I have mainly centred on Darwin's development of the principle of natural selection. But as Hodgson (1993) argues, this is only one of several vital components of his theory of evolution, another being the principle of variation and divergence. Based on Schweber (1980), Hodgson (1993) argues that Babbage is a likely source for this principle. Again, consistent with the present argument, Hodgson (1993) notes that Darwin and Babbage knew each other well and were members of the same scientific and social circles (Hodgson's statement is based on Schweber 1977b).
12 Chapter Seven of *Origin* introduces habits and instincts. Since instinct, for Darwin, referred to several distinct forms of 'mental actions' he declined to define the term, but suggested that '... every one understands what is meant, when it is said that instinct impels the cuckoo to migrate and to lay her eggs in other birds' nests' (Darwin [1859] 1985, p. 234). In the preceding chapters Darwin had variously referred to inherited modifications and cumulated differences in structure. Darwin considered the natural selection of instinct as a possibility that could co-exist with the inheritance of instinct and habit.

13 Hodgson (1993, p. 72) notes: 'We should not expect to find the theory of natural selection in the writings of Smith or Malthus, but this does not rule out the role of their ideas in providing crucial insights for the father of modern biology. And recognition of the abductive sparks moving from political economy to biology does not, of course, deny the importance of the medium of cultural and political environment at the time. Indeed, the oxygen is as vital as the spark itself.'

14 The term 'replicator' comes from Dawkins (1976) and Hull (1980). Dawkins (1976) proposed the term 'vehicle', whereas Hull (1980) preferred 'interactor'. I prefer the term 'interactor'. Apart from semantic connotation, the reason is that Hull (1980) introduced the term 'interactor' as a populational notion whereas Dawkins (1976) limited his 'vehicles' to development. The terms 'interactor' and 'replicator' have been proposed as useful for advancing a general selection theory encompassing biology, immunology and behaviour (Hull, Langman and Glenn 2001) as well as economics (Knudsen 2001).

15 This may as Hull (1990) suggests help reduce confusion over the nature of selection that comes from unclear terminology. Since the identification of an appropriate replicator and interactor demands detailed consideration of the implied selection process this exercise will provide an important step towards a tight evolutionary explanation.

16 According to the neo-Darwinian model, information flow from genes to soma cells is prevented by Weismann's barrier, the molecular barrier that prevents Lamarckian inheritance. In its modern form Weismann's barrier translates into the central dogma which states that information can flow from DNA and RNA to proteins but never in the reverse direction, that is, this defines the following relation between genes and soma cells: DNA <-> RNA -> protein.

17 Campbell ([1974] 1987, p. 56) insists that in all processes of knowledge growth, 'a blind-variation-selective-retention process is fundamental to all inductive achievements, to all genuine increases in knowledge, to all increases in fit of system to environment'. I have only recently come to realise that Campbell's ([1974] 1987) views on this point, as well as on the general solution, in terms of vicarious selection, relate to the obvious Lamarckian nature of much evolution in the social realm. My own views on these points concur with Campbell's ([1974] 1987) but I hope to add detail to his ideas in terms of the principle of local emulative selection proposed in previous work (Knudsen 2001), and used in the present.

18 To speak of 'Lamarckian' selection' to denote the inheritance of acquired traits requires a detailed specification of the causal relation among the constituent entities of social or economic evolution (Knudsen 2002). Since this required specification is omitted in most explanations of social or economic evolution, the use of the term 'Lamarckian' remains a mystery in most cases.

19 Binmore (1990, p. 17) notes that: 'In social science, usually only death eliminates the carriers of failed theories'. While still alive, these carries may nevertheless cease being replicating (perhaps no one cites, reads or listens to an unfortunate carrier of a failed theory). This seems to be very close to the fate of carriers of fit genes with low fitness: they don't necessarily die but on average they replicate less.

20 Schultz (1994) compares Kuhn, Bloom and Derrida and, consistent with the model of science proposed in the present chapter, argues that the cultural code, acting like the genetic DNA code in replication of science, poetry and philosophy, goes beyond the conscious. In Kuhn, Bloom and Derrida (as well as in Foucault, Baudrillard and Lyotard), Schultz (1994) finds arguments that the creative act in science, philosophy and poetry stems from a level of knowledge that goes beyond the individual's conscious reach. Thus, according to Schultz (1994, p. 462), 'The whole cultural form functions like a genetic code to reproduce the pattern – the matrix – of the discipline, yet a clear idea of the genetic code was never present to the creator as an abstract set of instructions for creation.' This chapter agrees on the conception of implicit knowledge as the defining characteristic of a discipline's matrix but differs in the conception of the level at which implicit knowledge is shared. Whereas Schultz (1994) insists the whole cultural form functions like a replicating genetic code, I suggest implicit knowledge is carried by the individual (the interactor) and replicated within face-to-face groups. That is, I suggest the 'whole cultural' form is seen as a (multi-dimensional)

distribution of replicating ideas and define the face-to-face group as the level at which implicit knowledge is replicated.

21 The professional identity is a particular aspect of the scientist's social identity. See Knudsen (2002) for an argument why the *interactor* in explanations of social evolution must be defined in terms of *social identity*.

22 This definition is consistent with Durham's (1991) definition of meme selection with culture substituted for scientific knowledge and meme substituted for individual scientific ideas.

23 Since most human acquisition of knowledge appears anything but 'blind', Campbell caused himself a lot of unnecessary problems by his use of 'blind'. The nested model proposed in this chapter is consistent with Campbell in the sense that it maintains an underlying process of implicit knowledge acquisition underlies any apparently explicit expressions of knowledge.

24 In *Origin* Darwin first presents domestic selection unconsciously done by the human breeder of dogs, plants and the like, a topic which would face less resistance than natural selection which he introduced by analogy. It is clear that Darwin focuses on unconscious effects of domestic selection of breeds of animals but he also indicates analogy between the evolution of animal breeds and human language: 'But, in fact, a breed, like a dialect of language, can hardly be said to have had a definite origin. A man preserves and breeds from an individual with some slight deviation in structure, or takes more care than usual in matching his best animals and thus improves them, and the improved animals slowly spread in the immediate neighbourhood.' Thus local emulative selection in the evolution of science involves: (1) an analogous principle of unconscious selection that works by emulation of the thought structure of scientists belonging to a particular discipline; and (2) the tendency for changes in thought structure will spread slowly in the local neighbourhood of scientists. There may for altogether different reasons than emulation of thought, such as status, appear a tendency to seek the company of excellent scientists. In this case the principle of local emulative selection is analogous to Darwin's breeding argument.

25 Also churches and military units where meetings take place among people in relatively stable face-to-face groups with relatively high frequency are good candidates for circumstances that should favour the evolution of implicit knowledge by local emulative selection.

26 This can be viewed as further specification of Cohen and Levinthal's (1990) notion of 'absorptive capacity'.

## REFERENCES

Binmore, Ken (1990), *Essays on the Foundations of Game Theory*, Cambridge, MA: Basil Blackwell.

Bloor, David (1976), *Knowledge and Social Imagery,* London: Routledge & Kegan Paul.

Bock, K. and Z. M. Griffin (2000), 'The Persistence of Structural Priming: Transient Activation or Implicit Learning?', *Journal of Experimental Psychological Genetics,* **129** (2): 177–92.

Campbell, D. T. (1969), 'Variation and Selective Retention in Socio-Cultural Evolution', *General Systems*, **16**: 69–85.

Campbell, Donald T. ([1974] 1987), 'Evolutionary Epistemology', in Gerard Radnitzky and William W. Bartley III (eds), *Evolutionary Epistemology, Theory of Rationality, and the Sociology of Knowledge*, La Salle, IL: Open Court, pp. 47–89.

Cohen, W. M. and D. A. Levinthal (1990), 'Absorptive Capacity: A New Perspective on Learning and Innovation', *Administrative Science Quarterly*, **35** (1): 128–52.

Collingwood, R. G. (1956), *The Idea of History,* Oxford: Oxford University Press.

Darwin, Charles ([1859] 1985), *On the Origin of Species by Means of Natural Selection, or the Preservation of Favoured Races in the Struggle for Life,* Harmondsworth: Penguin Books (first published by John Murray, London).

Darwin, Charles ([1887] 1969), *The Autobiography of Charles Darwin,* New York, NY: Norton.

Dawkins, Richard (1976), *The Selfish Gene,* Oxford: Oxford University Press.

Durham, William. H. (1991), *Coevolution: Genes, Culture, and Human Diversity,* Stanford, CA: Stanford University Press.

Eraut, M. (2000), 'Non-Formal Learning and Tacit Knowledge in Professional Work', *British Journal of Educational Psychology*, **70** (1): 113–36

Hodgson, Geoffrey M. (1993), *Economics and Evolution. Bringing Life Back into Economics,* Cambridge: Polity Press.

Hodgson, Geoffrey M. (1997), 'The Ubiquity of Habits and Rules', *Cambridge Journal of Economics*, **71** (6): 663–84.

Hull, David L. (1990), *Science as a Process,* Chicago: The University of Chicago Press.

Hull, D. L., R. E. Langman and S. S. Glenn (2001), 'A General Account of Selection: Biology, Immunology, and Behavior', *Behavioral and Brain Sciences*, **24** (3): 511–28.

Jacob, François (1985), *Mulighedernes Spil – Om Det Levendes Mangfoldighed (Le Jeu des Possible),* Copenhagen: Hekla.

Kentridge, R. W. and C. A. Heywood (2000), 'Metacognition and Awareness', *Conscious Cognition,* **9** (2): 308–12.

Kitcher, Philip (1985), *Vaulting Ambition: Sociobiology and the Quest for Human Nature*, Cambridge, MA: MIT Press.

Knudsen, T. (2001), 'Nesting Lamarckism Within Darwinian Explanations: Necessity in Economics and Possibility in Biology?', in J. Nightingale and J. Laurent (eds), *Darwinism and Evolutionary Economics*, Cheltenham: Edward Elgar, pp. 121–59.

Knudsen, T. (2002), 'The Significance of Tacit Knowledge in the Evolution of Human Language', *Selection,* **3** (1): 93–112.

Kohn, David (ed.) (1985), *The Darwinian Heritage,* Princeton, NJ: Princeton University Press.

Liebermann, M. D. (2000), 'Intuition: A Social Cognitive Neuroscience Approach', *Psychological Bulletin*, **126** (1): 109–37.

March, J. G. (1991), 'Exploration and Exploitation in Organizational Learning', *Organization Science*, **2** (1): 71–87.

March, J. G. (1999), 'The Evolution of Evolution', in James G. March (ed.), *The Pursuit of Organizational Intelligence*, Malden, MA: Basil Blackwell.

March, James G. and Herbert A. Simon (1958), *Organizations,* New York, NY: Wiley.

Maynard Smith, John and Eörs Szathmáry (1999), *The Origins of Life: From the Birth of Life to the Origin of Language,* Oxford: Oxford University Press.

Mayr, Ernst (1982), *The Growth of Biological Thought: Diversity, Evolution, and Inheritance,* Cambridge, MA: Belknap Press.

Mirowski, Philip (ed.) (1994), *Natural Images In Economic Thought,* Cambridge: Cambridge University Press.

Nelson, R. R. (1995), 'Recent Evolutionary Theorizing about Economic Change', *Journal of Economic Literature*, **33** (1): 48–90.

Nelson, Richard R. and Sidney G. Winter (1982), *An Evolutionary Theory of Economic Change,* Cambridge, MA: Harvard University Press.

Nisbett, R. E. and T. D. Wilson (1977), 'Telling More Than We Know: Verbal Reports on Mental Processes', *Psychological Review*, **84**: 231–59.

Nonaka, Ikujiro and Hirotaka Takeuchi (1995), *The Knowledge Creating Company,* Oxford: Oxford University Press.

Polanyi, Michael (1957), *Personal Knowledge: Towards a Post-Critical Philosophy,* London: Routledge & Kegan Paul.

Polanyi, Michael (1966), *The Tacit Dimension*, Garden City, NY: Doubleday.

Popper, Karl R. (1979), *Objective Knowledge: An Evolutionary Approach,* Revised edition, Oxford: Oxford University Press.

Reason, James (1990), *Human Error*, Cambridge: Cambridge University Press.

Reber, Arthur S. (1993), *Implicit Learning and Tacit Knowledge: An Essay on the Cognitive Unconscious,* Oxford: Oxford Unversity Press.

Richards, Robert J. (1987), *Darwin and the Emergence of Evolutionary Theories of Mind and Behavior,* Chicago, IL: The University of Chicago Press.

Rosenberg, Alex (1994), 'Does Evolutionary Theory Give Comfort or Inspiration to Economics?', in Philip Mirowski (ed.), *Natural Images In Economic Thought*, Cambridge: Cambridge University Press, pp. 384–407.

Ryle, Gilbert (1971), 'Knowing How and Knowing That', reprinted from *Proceedings of the Aristotelan Society*, vol. XLVI, 1946, in Gilbert Ryle, *Collected Papers Vol. 2*, London: Hutchinson, pp. 451–64.

Schultz, William R. (1994), *Genetic Codes of Culture: The Deconstruction of Tradition by Kuhn, Bloom, and Derrida,* New York, NY: Garland Publishing.

Schweber, S. (1977a), 'Darwin and the Political Economists: Divergence of Character', *Journal of the History of Biology*, **10**: 231–316.

Schweber, S. (1977b), 'The Origin of the *Origin* Revisited', *Journal of the History of Biology*, **10**: 229–316.

Schweber, S. (1980) 'Darwin and the Political Economists', *Journal of the History of Biology*, **13:** 195–289.

Sternberg, Robert J. and Joseph A. Horvath (1999), *Tacit Knowledge In Professional Practice,* Mahwah, NJ: Lawrence Ehrlbaum Associates.

Toulmin, Stephen E. (1972), *Human Understanding,* Princeton, NJ: Princeton University Press.

Turner, Stephen (1994), *The Social Theory of Practices: Tradition, Tacit Knowledge, and Presuppositions,* Chicago, IL: University of Chicago Press.

Wittgenstein, Ludwig (1969), *On Certainty,* New York, NY: J. & J. Harper.

Woltz, D. J., M. K. Gardner and B. G. Bell (2000), 'Negative Transfer Errors In Sequential Cognitive Skills: Strong-but-Wrong Sequence Application', *Journal of Experimental Psychology, Learning, Memory and Cognition,* **26** (3): 601–25.

Worster, Donald (1994), *Nature's Ecology: A History of Ecological Ideas,* Second edition, Cambridge: Cambridge University Press.

# 6. Science and spontaneously formed institutions: an Austrian School approach

**Laurence S. Moss**

## 1. INTRODUCTION

In a recent paper that appeared in the *Journal of Economic Methodology*, Esther-Mirjam Sent surveyed the economic literature about science and its broader organisational features. While concluding that there is 'little consensus over what an economics of science would be like' (Sent 1999: 114), she hoped her discussion might pave the way toward a more 'synergetic approach to "science" and the "economy"' (Sent 1999: 115). Sent did not consider whether or not the modern Austrian School might have something to offer on this topic, and this omission is what encouraged me to prepare this chapter. What might an Austrian School inspired approach to the economics of science look like and how might it proceed so as to offer insight and understanding about science, its historical development and its organisation?

The 'spontaneous order' approach that I offer here has obvious links to the work of Carl Menger in the 19th century and to Friedrich von Hayek and the modern Austrian School tradition in the 20th century. Austrians have expressed much interest in what we called 'spontaneous social formations', that is, those rules, habits and customs that arise slowly and gradually over time but function to promote ends that were really not envisioned by those who 'fell into line' behind the rules as they originally emerged. Austrian School economists make much of these spontaneously formed institutions and the difficulty of identifying them with any precision. In some accounts, Austrians remind politicians and exuberant legislators to slow down their interventions. Many of the extant rules, habits or customs now prevailing in the market promote economic coordination and material prosperity in ways that remain largely unknown and remarkably unappreciated. To meddle, prohibit or in other ways weaken their operation may unlock a Pandora's box of horrifying consequences. Austrian economists remind us that much of the suffering of the 20th century was caused or at least exaggerated by the pretensions of intellectuals who thought that they could plan an economic system in the same way as a potter conjures up a ceramic bowl out of clay mud (Hayek [1988] 1989).

But it is not only politicians and legislators that ignore or try to change the institutions of society. Scientists sometimes violate institutions within their scientific communities in their quest for a discovery or killer breakthough to make them famous. It is not only that they think in a novel way about a problem in science and break out of the established analysis, it is also that they might intentionally jump the queue, steal a look at another's private research, and even lie, cheat and so forth, and in this way get to the finish line first to claim the prize! When they act in this way, they set off rancorous debates within their ranks; it is these debates that provide us with a clue about how to develop an economics of science.

A genuine economics of science might start with the identification of one or more Austrian School style institutions and then explore the mechanisms by which they operate to promote those ends which were never intended by those who conform to or replicate those institutions in their daily professional behaviour. Typically these institutions are not specified in any explicit detail. Spontaneous institutions are most often implicit in the language scientists use, in their likes and dislikes, and are more a matter of habit and custom than a dramatic set of rules posted on the door to a laboratory. That is why we take a sharp interest in certain episodes in the history of science. Those who chronicle these historical 'moments' often tell of a clash between scientists which accompanies an intense competition toward some final breakthrough discovery. These rancorous episodes bring out and expose the broader institutions at work within the community of scientists. It is not uncommon for warring scientists to accuse each other of violating certain long-established rules and protocols. From these accounts we can infer some of the institutions at work and then speculate about how they function to either promote or retard the scientific work underway.

To the extent that the Austrian spontaneous order approach is dubbed 'economics' (and I would insist that under a definition of economics which includes the study of coordinative mechanisms this is indeed economics) we have the scaffolding for a genuine economics of science. In the next section of this chapter, I review the spontaneous order approach as it has been presented in the Austrian School literature. I emphasise that it is the Menger-Lachmann strand of thinking that I follow, which supports a distinction between the origin and formation of the rules that guide conduct and the normative evaluation of how those rules actually have functioned in particular historical settings. Institutions can promote results that we value, even if they originated in conscious deliberation and were the outcome of parliamentary edicts.[1]

In the third section of the chapter, I review Thomas S. Kuhn's influential ideas about a scientific community and its operation. From here I analyse two important episodes in the history of science: first, the discovery of insulin in Toronto in the 1920s; and second, the famous Watson account of the discovery of the structure of the DNA molecule in the 1950s. Again, my purpose in surveying these historical episodes is to identify several examples of institutions that guided the scientific community in the 20th century and to assess to what extent these institutions functioned to promote or retard the development of scientific inquiry. I shall demonstrate that definitions of scurrilous behaviour among scientists often turned on actual or alleged rule-breaking behaviour. One special question that pops out of the analysis I offer here is whether the 20th century's move toward team-based research required the elimination of several long-standing customs and rules which guided the scientific community. If we answer this question in the affirmative, then we have discovered still another instance where rules of conduct of long and established provenance may have persisted into a time when they have ceased to be useful or at least are no longer perceived as being useful. This, in my opinion, is a good start for an Austrian-based economics of science. I emphasise this point again in the conclusion of this chapter.

## 2. SPONTANEOUS SOCIAL FORMATIONS

Let me begin with a phrase from Norman Barry's 1982 survey piece. According to Barry, there are 'things of general benefit in a social system [that] are the product of spontaneous forces that are beyond the direct control of man' (Barry 1982: 7). What are these 'things' of general benefit that appear in a social system and how do they operate to promote some ultimate end or objective that we consider 'beneficial'? The examples that Barry offers range from the coordinative role played by competitive market prices to the familiar example of the common law legal rules, and include economic institutions that function to economise on information and lower transaction costs, such as the money commodity itself (Barry 1982).

Barry drew on a historical tradition that his mentor, Hayek, also found to be of great importance. Hayek, it will be remembered, criticised the sharp distinction in the social science literature of his day between 'things' that are in some sense 'natural phenomena' like the weather, earthquakes or natural barriers like mountains, and other 'things' that, unlike the weather, earthquakes and mountain ranges, are created by men and women. These 'things' are not 'artificially created' as might be a set of statutes ground out by a legislative body regulating, say, crowd behaviour on a particular outdoor civic space. Rather, both Barry and Hayek were interested in the origins of certain rules that arise of out of human interaction and that are replicated by daily behaviour but do not arise out of any intentional plan or legislative body in which minds meet, deliberate and decide. Also, these rules, habits and customs are far removed from human instinct – there may be nothing biological about them. Still, despite their humble and non-deliberative origins, these 'institutional rules' miraculously function to promote larger societal goals or designs.

Hayek harked back to an older tradition in the social sciences – the Scottish Enlightenment of the 18th century. The Scottish thinkers initiated the search for the organising rules of social and economic organisation. One member of that group of pioneering social scientists, Adam Ferguson, summarised the very idea that Hayek took over and developed further. Ferguson's exact passage in which he called attention to those 'things' that are created by humans, but not rationally or deliberatively, is as follows:

> Every step and every movement of the multitude, even in what are termed enlightened ages, are made with equal blindness to the future; and nations stumble upon establishments, which are indeed the result of human action, but not the execution of any design. (Cited in Barry 1982: 24)

What Ferguson called 'establishments' and Barry and Hayek termed 'things' we shall simply call rules, habits, customs or more generally 'institutions'.

In this category, I include the following: choosing business and research partners; communicating, negotiating and entering into contracts; passing the ownership title of land from one individual or family to the next; transferring titles of ownership over money deposits in one location to a new owner while positioned far away in another location; and, perhaps most important and fundamental of all, designating certain individuals as having the power to authenticate matters needed to support status relationships in society (Hayek [1988] 1989, p. 12; Searle 1995, pp. 79–112). The modern Austrian School has not emphasised the role that status relationships and authentication procedures play in coordinating human action but much can be gained by considering these issues as well.

Authentication is a valuable service that is mostly overlooked. Let us take a short detour and explain the purpose of authentication services. It matters in most villages, towns and cities around the world whether two cohabiting individuals are married or not. Suppose Mary was living with John but not married to him. John suddenly breathes his last breath. In many legal systems, if Mary was his wife she would immediately become the legal owner of his property. If not, she might have to leave the house so that the lawful heirs could move in. The status of 'married' is more than just a self-proclaimed puff title like 'best pizza maker in Boston', rather, it triggers under certain well-known contingencies exact and predictable patterns of behaviour on the part of many others.[2] The presence of relationships such as these also serves to make private planning more effective and human interaction less uncertain. Similarly, when a graduate student becomes a 'Doctor of Science' or, as is common in Europe, is designated fit to teach in university, a new set of duties and entitlements opens up for that former student. Her or his status changes and with it the behaviour of others who relate to him or her in new ways.

I could go on with other examples. My point is a simple one. In order for status relationships of these types to come into existence and function to help the rest of us accomplish our ends and goals, there exists a whole host of people who themselves hold various status positions in society, and who must via customary acts behave in patterned ways so as to authenticate the status of others. Once a particular individual J is designated as having a status S, others in society will be called upon from time to time to authenticate that status. These others will be asked to answer such questions as, 'Does the doctor with whom I have contracted to perform a certain surgery on my body actually have the status of being "board certified" such as appears on his letterhead, or is he a fraudulent doctor?' Or, perhaps, 'Does this man Mr X who claims to have witnessed Ms Y's voluntary signature on this important document really have the status of being a notary in London? Whom can I ask in London to authenticate the questioned fact that Mr X is indeed the notary that he pretends to be when he authenticates the signatures on this document?' The authenticator might demonstrate his or her status by signalling with a special seal or stamp of his or her own. And what about the task of distinguishing honest from counterfeit money?

Modern Austrians recite Menger's remarkable account of the origin of money as the premiere example of a spontaneously formed institution. The analysis can be expanded by taking notice of authentication activities directly linked to the emergence of money itself. Let us see how this can be done. According to Menger (and as later elaborated by von Mises ([1949] 1963, pp. 398–416) individuals stumble upon 'indirect exchange' as a way of accomplishing their plans. In an indirect exchange, the rational trader accepts a commodity that he or she does not value directly (that is, for its 'intrinsic qualitites') but only indirectly. This commodity that he or she has accepted in exchange is perceived to be 'marketable', and therefore he or she calculates that in the (near) future he or she will find a third party that will accept it in exchange and give to the trader something which he or she needs more urgently. Interestingly, because the exchange items are perceived as marketable, they in turn actually become more marketable as others learn about this, and this in turn enhances their reputation still more. This is a positive feedback effect mechanism at work (Arthur 1994, pp. 1–12). Menger was one of the first to point out the phenomenon and emphasise its favourable consequences for economic life, using it to explain the origin of money.

Money is the classic example of a social institution that, arising spontaneously, helps a community of strangers to coordinate their respective activities and build a successful economy. The Mengerian account explains how the mere self-interest of traders in finding the most marketable goods creates a situation in which some goods do achieve the reputation of 'most marketable'. Money-like goods become the temporary abodes of purchasing power that help traders sell without committing to a purchase. With the benefit of the passing of time and new information, traders can plan their next purchase more rationally. Historically, gold and silver have been easy candidates for the status of money because of their portability and distinctive appearances. Gold and silver became the money-commodities in many parts of the world. But the purity of the gold and the silver offered in exchange and their exact weight are always matters for conjecture and argument. Even state-coining with the stamp of the ruler is not always enough to reassure traders that what they are accepting in exchange is likely to be acceptable to the next seller later on. The coins that are offered may have been clipped or may be worthless impostors. And so an army of authenticating specialists emerges to identify impostor money from real money. Now the authenticators are called upon to authenticate a fact about the world – 'Is this coin made of pure silver and how much does it weigh?' In our earlier examples authenticators were called upon to decide if an individual possessed a certain status – 'Is this man a notary and capable of performing the ritual of authenticating voluntary signatures?' 'Is this woman married?' In either case, the proclamation of the specialist sets in motion a plethora of renewed understandings and expectations that are at the heart of social and economic life.

Commerce relies on the testimony of the expert, an authentication specialist of some sort, to proclaim that 'This coin is genuine and contains so much weight of silver', or 'This woman is a board-certified surgeon', and so on. Without this army of authenticating experts the marketability of the money-commodity would be substantially weakened, the value or the doctor's services lessened, as doubts and suspicions add costs to the simplest acts of trade and exchange. Without the availability of low-cost authentication services of experts, the positive feedback process which catapulted money into existence becomes much slower and perhaps is dampened to a snail's pace if not eliminated completely (Hodgson 1993, p. 117).

In sum, whatever their origins – humble or extraordinary – status relationships are created, perpetuated and authenticated by a vast number of individuals who perform a valuable and often overlooked service in any social and economic system. Their methods of communication with each other and with their audience seeking information about status and its authentication are part of the institutions of modern society. They are the 'establishments' to which Ferguson alludes. They are manifestations of the institutions that make human action more effective, the division of labour more extensive and the region in which the institutions are perpetuated wealthier. Without institutions of this sort, individuals would have difficulty communicating at great distances, and the coordination of their economic activities to their mutual advantage would be difficult, if not impossible.

It is clear that the founder of the Austrian School, Carl Menger, was genuinely interested in institutions that arise in the intriguing manner that Ferguson described, that is, with humans acting largely blind to the process but somehow groping and stumbling upon rules, routines, habits and customs that are later designated 'enlightened' and worth preserving. Menger termed these practices 'organic institutions' and pointed to the origin of money and the common law as his paradigmatic examples. But the set of all beneficial institutions was not exhausted by the 'organic institutions' that Menger and later Austrians identified. There were many other institutions with a different provenance that interested Menger and that were worth preserving. Indeed, the search is still on.

Before continuing the search in the next section of this chapter, I wish to point out that there are also beneficial institutions that arise in an entirely different manner, that is, as a result of planning and deliberation on the part of humans. Menger termed these other rules of conduct 'reasoned rules of conduct'. Like the unintentionally created rules, they also serve with honour and distinction in promoting important goals and benefits that were a precise part of their original intention. Comparing the spontaneous social formations with the deliberately engineered institutions, Menger was crystal-clear that one set of institutions was not in any way more privileged for sainthood than the other set. Consider this passage from the English translation of Menger's 1883 work, *Problems of Economics and Sociology*, in which Menger makes it clear that law can originate both 'organically and also as the result of human intelligence':

> [L]aw is not always the result of an (intended) *common will* directed toward establishing it and toward the furthering of human well-being. Originally [law] was not this at all. This [spontaneous origin of the law] by no means excludes the

genesis of law as the result of human intelligence. (Menger [1883] 1963, p. 230 note)

Other Austrian writers, including Ludwig Lachmann, followed Menger closely on this point. According to Mufit Sabooglu, in Lachmann's work a clear distinction is made and maintained between the origin of a beneficial rule or custom and the secondary requirement that the user of the custom 'must not be conscious of its value'. For Lachmann certain institutions will work quite well regardless of whether or not the user is conscious of their value and regardless of whether they originated in custom or in the parliament (Sabooglu 1996: 362).[3]

## 3. THE SCIENTIFIC COMMUNITY

Thomas S. Kuhn called it 'the essential tension' (Kuhn 1977). On the one hand, scientific research proceeds in a social framework of well-established traditions and habitual routine practices. Those who come on the scene to undermine or cast aside those practices will find their careers in science challenged; seasoned mavericks and dissenters might never receive their terminal degree. On the other hand, there are episodes in the history of any science when certain individuals emerge and shatter old beliefs with new ways of thinking about a problem or reasoning and proceeding toward the understanding of some phenomenon. Most of those who promulgate the revolutions in thinking themselves struggle in both worlds, since it is helpful if the innovator is also capable of doing 'normal science' and thinking through problems in the way expected of him or her by teachers and by elders. This serves to replicate the established institutions of science within a particular scientific community.

According to Kuhn, 'normal research, even the best of it, is a highly convergent activity based firmly upon a settled consensus acquired from scientific education and reinforced by subsequent life in the profession' (Kuhn 1977, pp. 225–39). But the discoverer of new information and principles must sometimes be capable of breaking ranks with the authority of tradition and taking bold points of departure. Yes, indeed, there does seem to be a primordial tension between the two processes, each important yet destined to come into conflict from time to time. Rule-making and rule-breaking behaviour are both the stuff of scientific discovery and breakthrough experimentation.

The historiography of science often makes a related tension the centrepiece of the account offered. There is the proverbial Galileo grasping a new concept of the solar system and then confronting those who held on to the old-fashioned ideas of their day. New ideas are put on trial, along with the mavericks who propose them. In this and other examples we learn about the tension that exists between scientists and the governing institutions that guide them to replicate their past. By conforming to these traditions, the scientist passes down these patterns and conduct to the young and uninitiated. The innovator, however, sees things in a brand new way and sometimes breaks with customs and rituals, thereby pushing science in new directions. Often these same individuals not only break with the older moulds of thought but also challenge the authenticators by overturning customs. Rather than perpetuate established institutions, these innovators challenge and criticise the legacies of the past. It is in these moments of scientific tension that the prevailing and governing institutions are suddenly highlighted and stripped naked. There they are. The spontaneous institutions that in many ways operated with subtlety and by way of shared meanings and implicit gestures are now at the centre of debate. An Austrian-style economics of science needs first of all to identify these institutions, articulate some of their meanings and interrogate the manner in which they function. This can be done, and that is why we must now turn our attention to two dramatic cases of scientific discovery in the 20th century.

## 4. THE DISCOVERY OF INSULIN

Consider the discovery of insulin – 'one of the most important medical discoveries of the modern age' (Bliss 1982, p. 189). The Nobel Prize was awarded to both Dr Frederick Grant Banting and Dr John James Rickard Macleod in October of 1923. The award of the prize led to numerous lecture invitations in which Banting and Macleod, travelling on separate circuits, offered conflicting accounts of the discovery of insulin. Banting was particularly ferocious in his treatment of Macleod's contributions to the scientific breakthrough. According to Banting, Macleod 'was never to be trusted', 'sought at every possible opportunity to advance himself', and was always 'grasping, selfish, deceptive, self-seeking and empty of truth, yet he was clever as a speaker and writer' (Bliss 1982, p. 202). The obvious conclusion is that Banting did not consider Macleod's contribution worth recognition and resented having to share the Nobel Prize with this imposter. The dispute, as chronicled by Bliss's award-winning account of the affair, brings out the role that institutions and rituals of accepted conduct played in this 1920s drama (Bliss 1982).

Macleod served as a Professor of Physiology at the University of Toronto and was an internationally recognised expert in carbohydrate metabolism. The University named him to serve as research director of their newly created laboratory associated with the medical school. It was in his capacity as research director that Macleod responded to an autumn 1920 job application by a University of Toronto graduate to perform a series of experiments on the internal secretion of the pancreas, under the auspices of the University. The applicant, Banting, wanted Macleod to hire him as a salaried researcher in the newly created laboratory.

At that time, Banting was a practising doctor with a specialisation in surgery. His London practice was without much financial success. For several reasons both personal and professional, Banting wanted to get away from London and try his hand at a different vocation. He wished to return to Toronto by first obtaining an academic research position.

With only a cursory reading of the literature on pancreatic secretions, Banting had come up with an original experimental design that involved ligation of the pancreatic ducts in laboratory animals and then testing to see what extracts might indeed be present. The idea was to catch the extracts before they mixed with the other digestive materials, a feat that had never before been tried. This turned out to be the key to isolating insulin – the idea of capturing it before it mixed with other body fluids.

As laboratory director, Macleod was required to conduct an interview with research applicants. The interview involved a plethora of authenticating rituals and customs to which candidates had to submit before precious research materials and other scarce resources would be given over to them. The Macleod-Banting interview did not go well. The meeting took place on 7 November 1920 and Macleod recorded his best memory of that meeting less than two years later. It was clear that Banting had imagined a way of surgically isolating the pancreatic excretions before they mixed with the external secretions of the digestive process, and on this point Banting's proposed experiments interested Macleod. But Macleod, after acknowledging the creativity of the proposed approach, probed further to see if Banting had enough background to carry out the proposed research. Banting's knowledge of other related matters was unsatisfactory and in Macleod's mind scandalously incomplete.

Macleod recorded that 'Dr Banting had only a superficial text-book knowledge of the work that had been done on the effect of pancreatic extracts in diabetes' (Bliss 1982, p. 52). Indeed, it is not hard to imagine the incredulity of the research director to a 'young surgeon who had walked in virtually off the street, had no significant experience in physiological research, and was talking, haltingly, about a topic he knows about only from standard textbooks and one article.' Macleod's scepticism boiled over when he reminded the young Banting of the multitude of eminent scientists who had worked for years on the problem of the pancreas and had been unable to find or even prove that such an internal secretion existed. Macleod's negativism would be harboured in the mind of the young applicant and apparently fuel a lifetime of distorted memories and rage, culminating years later in the row about the propriety of naming Macleod a co-discoverer of insulin. Here was a screening mechanism of historic and longstanding use, which Banting considered useless and inappropriate when it was applied to his candidacy. Macleod applied the mechanism and emphatically shared his concerns with Banting. As a result, Banting formed a deep dislike of Macleod. This example reveals what the authentication mechanisms were like during the 1920s in the emerging university laboratories and how they were enforced.

After this initial interview in which Macleod questioned Banting's knowledge of previous research, he went ahead and agreed to put the young surgeon on the staff. Why did Macleod do this? Did he perceive the novelty and potential of Banting's proposed surgical procedure? Macleod's willingness to take the risk that the coherence of the surgical procedure would offset Banting's other failings was truly entrepreneurial. Macleod was careful to shape a research programme that would offset Banting's many academic deficiencies. We shall see that the laboratory director's contribution towards organising the research, appointing others to authenticate the results and even publicising the findings among the pharmaceutical corporations of the day was also an important factor in the 'discovery' of insulin. The Nobel Prize committee clearly recognised the importance of team-based science.

Banting was provided with some space in the laboratory and was also assigned a young science major, Charles Best, to do the chemical tests that were necessary for the project to succeed. This much secured, Macleod went off to enjoy a vacation in Scotland during the summer of 1921.

Upon Macleod's return, the Banting-Best team reported remarkable results. The extract of pancreas that they had made from duct-ligated dogs had the effect of lowering both the blood sugar level and other symptoms in diabetic dogs (Bliss 1982, p. 2). Seeing obvious medical promise in the research results, Macleod quickly expanded the Toronto team of researchers to determine the physiological action of insulin. The expanded team included other first-rate Canadian scientists such as James Bertram Collip.

The idea that discovery in science is often a team effort seems to have been at the heart of the resulting priority controversy. It is important to emphasise this point. There was never any dispute that the credit for the duct-ligation experiments should go entirely to Banting and his co-worker Best. What was in dispute was whether the subsequent isolation and purification of insulin were essential to the discovery of insulin.

Macleod's conduct in the whole affair was exemplary. Indeed, Macleod had declined Banting and Best's offer 'to add his name to their first paper, published in the February 1922 *Journal of Laboratory and Clinical Medicine*' in which the findings of the duct-ligation experiments were first announced (Bliss, 1982, p. 197). It was the custom for laboratory research directors in the 1920s to have their names added to the papers written by others who engaged in work under their direction, a practice which remains today in most research centres around the world. However, Macleod asked that his name not be included since he did not claim credit for this particular work involving the ligation experiments.

If the Nobel Prize had been awarded solely for the findings of the duct-ligation experiment, then the Banting-Best team would have received the Nobel Prize and Macleod's name would not have been mentioned. Had this happened there never would have been any authorship controversy to grip the attention of the Canadian media. But it was the organisation of the entire project and process that ended up with the purified insulin that constituted the discovery in the eyes of the Nobel Prize committee. The lonely inventor model that might have applied to science in the 17th, 18th and early 19th centuries had long given way to the modern research laboratory and the concept of team-based research (Mowery and Rosenberg 1998, pp. 11–46). Macleod's administrative ability was perfect for the task that needed to be done. Banting's credentials suggest that he could not have done it alone. Macleod's skilful organisation of the research project undoubtedly contributed to the success of the project.

It turned out that the Nobel Prize was awarded for the discovery of insulin in its purified form. This included purification of the extract, making Macleod's contributions and insights very important to the final success of the project. According to Macleod, purification involved a large amount of research, including 'the investigation of rabbits, of glycogen formation, of acetone excretion, of respiratory quotients' and other matters 'done by members of a team working under his direction' (Bliss, 1982 p. 198). Bliss concludes that 'Macleod believed ... that Banting and Best would not have come close to insulin without his and then Collip's help' (Bliss, 1982, p. 202). In summary, Macleod believed that the discovery of insulin 'depended on the conjoint efforts of several investigators working under [Macleod's] direction, of which Dr Banting was one' (cited in Bliss, 1982, p. 198). And that is why the Nobel Committee awarded the Nobel Prize to both Banting and Macleod – justice was done!

This would have been the end of the story had not Banting decided to deride Macleod and question the propriety of placing the research director's name on that prize which in his view belonged solely to himself and his laboratory assistant. It was the outspoken Banting who caught the attention of the media, and his accounts of the discovery that appeared in Baron's and other sources succeeded in demeaning Macleod's role in the process and making Macleod's personal life miserable (Bliss 1982, p. 202). Banting's accounts of past meetings and conversations with Macleod portray the laboratory director as unhelpful, even spiteful. But the evidence suggests that Banting's accounts were biased, mean-spirited, and his memory selective. According to Bliss (1982, p. 50), 'Banting was not a precise and reliable guide to the events in which he participated.' Banting displayed no understanding or appreciation of team-based research. His early profession as a virtuoso surgeon may have imbued him with a 'cowboy' mentality about research itself. Such views were out of step with the imperatives of team-based research and the modern research laboratory that was rapidly making headway in the first quarter of the 20th century.

Banting's appreciation of authorship reflects a tradition in decline. It was no longer possible for the lonely individual burning the midnight oil to come up with all of the significant breakthroughs that change the ways we live and work. The customs and habits of an earlier period in the development of science were undergoing moderation and alteration as the modern research laboratory emerged with its requirement of divided specialities and labour coordinated by a wise research director.

The controversy gives us some insight into the essential tension between the institutions of normal science and the perceptions of those who come in to upset the verities of established science. Banting's inexperience with prior research yet dedication to the tasks ahead may actually have contributed to his success. Mavericks often do stumble upon insights and inventions, precisely because they are not burdened by certain frames of reference that might blind them to that which needs to be noticed. The historical literature is filled with accounts of inventors who came to their projects without any rigorous academic training regimen. Still, it is understandable why a research laboratory director in Macleod's position would have a duty to authenticate an applicant's suitability for a job. By the prevailing academic customs and routines, Banting was not an expert on the physiology of diabetes. He needed help. Indeed, had Macleod adhered to the authentication rituals of his day, Banting might not have been hired at all.

But Banting was appointed by Macleod. It seems that Macleod took a great risk and suffered deeply in his later life from the personal attacks. Still, the 'wrong' man ended doing the right job. The duct-ligation experiments worked and insulin was discovered. Rather than thank Macleod for his courageousness and willingness to depart from custom, Banting developed a deep hatred of Macleod. Was the hatred due to the fact that Macleod had followed the established rituals? With suspicions as the order of the day, the prevailing customs and habits of a well-trained research director now received twisted reinterpretations by the angry Banting. Consider another incident involving Macleod's laboratory.

On 30 December 1921, at the American Physiology Society meetings held at Yale University, the results of that summer of important laboratory research were to be reported in the form of a joint paper entitled 'The Beneficial Influences of Certain Pancreatic Extracts on Pancreatic Diabetes'. The co-authors of this part of the research included Macleod and Banting. Banting spoke of the research but his talk 'fell flat' and failed to convince the army of expert sceptics that he had finally succeeded in proving that there was indeed an 'internal secretion of the pancreas' that lowered the blood sugar level (Bliss 1982, pp. 104–5). When all hope seemed lost, Macleod took centre stage and made a spirited case for this research, along with its expected benefits to humanity. His manner of delivery and the persuasiveness with which he recounted the evidence helped save the day and left the scientific audience with some understanding of what had been accomplished in his laboratory at the University of Toronto. Macleod turned this meeting around, and by the end he had accomplished his objective of bringing great prestige to his laboratory and attracting corporate interest in its scientific findings.

This intervention would become a second source of anger and resentment on Banting's part. Banting saw Macleod as intentionally stealing the spotlight and claiming the discovery as his own, despite the fact that in his subsequent memories of the Yale meetings, Banting readily admitted that he failed in effectively presenting his research (Bliss 1982, p. 107). The customs and duties of the laboratory director required a form of behaviour designed to bring credit to the Toronto centre and lay the foundations for the future direction of research money and even industry attention. As it would turn out, between 1923 and 1967, the University of Toronto would gross over $8 million in royalties from the various intellectual property rights associated with the discovery and purification of insulin. It was Macleod's job as laboratory director to lay the foundations for this later result. His intervention must be credited as an important milestone in the larger effort to finally commercialise insulin and provide a medicine for diabetics, first in North America and later around the world. Still, the customs and practices of the smoother and more articulate doctor received a sardonic interpretation from Banting.

The authority dispute between Banting and Macleod has had the unintended effect of revealing the norms and customs of the North American scientific community on the eve of one of the greatest medical discoveries of the last century. The episode can be mined for further examples and evidence of how the emergent norms and institutions of what some might call 'big science' typically functioned to give shape to scientific progress in the 20th century. Despite Banting's complaints about the process of attributing authorship to scientific achievement, it is clear that the process as a whole had not malfunctioned. Banting may not have understood that the 'cowboy' inventor model of scientific discovery was taking a back seat to the team-based research that would catalyse so many of the scientific accomplishments of the 20th century. It is also clear that when the imperative of teamwork is taken into account, Macleod's name rightfully belonged on that Nobel Prize along with Banting's. According to Bliss, the laboratory director carried out his duties with speed, conscious attention to detail and most effectively as well.

## 5. THE DISCOVERY OF THE STRUCTURE OF DNA

For an example of a case in which the customs and institutions of big science may have actually retarded innovation and discovery, we turn to James Dewey Watson's personal account of the discovery of the structure of DNA, for which he shared the Nobel Prize for Medicine and Physiology with Francis H. C. Crick and Maurice H. F. Wilkins. Again we have an example of team-based research and discovery. This time several of the prevailing institutions of big science acted as barriers to the breakthrough discovery. The ingenuity and some might say ruthlessness of the behaviour of the Watson team is what coordinated and accomplished the discovery effort. Here the goal was to win the Prize and the means used to accomplish this end might indeed have left something to be desired.

The young American, Watson, was sent to Copenhagen in the spring of 1951 to learn all he could learn about biochemistry. Watson was not at all stimulated by the Danish laboratory to which he had been assigned and his interests turned to other matters. Those other matters included a remarkable photograph of DNA Maurice Wilkins showed to a group of scientists in Naples. Wilkins had taken the photograph using the latest X-ray technology. Watson was in Naples at the time and happened to catch the lecture. Apparently, the Watson-Wilkins meeting was entirely fortuitous since Watson was in southern Italy in order to escape the cold northern European winter. Wilkins's X-ray image suggested that DNA was a crystalline substance and that there was much to learn about its structure in Wilkins's laboratory in England, and not much for Watson to learn by remaining in Denmark. Based on this experience, Watson 'knew that X-ray crystallography was the key to genetics' (Watson 1968, p. 43). Apparently, this was an important turning point in the history of the discovery of the double helix, and it had all come to Watson in something of a flash.

But how could Watson get from Denmark in northern Europe where he was sent to study biochemistry to the University of London where Wilkins was a laboratory director (Watson 1968, pp. 21–8)? The switch of venues was problematic because Watson was on a post-doctoral fellowship and the funding agency in Washington, DC most likely would not approve Watson's sudden decision to abandon biochemistry for what they would no doubt perceive as the unrelated field of X-ray photography. As a result, Watson deceived the Washington, DC bureaucracy by asking an English biochemist, Roy Markham, to agree to allow his name to be used as a straw so that Watson could be assigned to the Cavendish Laboratory in Cambridge and still receive his government funding. Watson made it look like he would be studying biochemistry in Markham's laboratory, but he made it clear to Markham that he would not be around at all to bother him. Since fraud – that is, the intentional misrepresentation of a material fact on which (in this case) a government agency is asked to rely when funding a post-doctoral student – constitutes the violation of a rather serious norm in academic settings of truthfulness and honesty, it seems that, had Watson not engaged in this chicanery, the Crick-Watson hypothesis which is regarded as one of the greatest discoveries of all time might not have taken place when it did. On the other hand, in Watson's defence, the importance of funding research on large-scale projects with government money brings in the bureaucracy which in so many ways is incapable of understanding the subtle linkages between biochemistry and X-ray photography.

This violation of norms was not the only one Watson was involved in. Francis Crick was at the Cavendish Laboratory with a brilliant mind and a talent for quickly 'seizing [other scientists'] facts and ... reduc[ing] them to coherent patterns' (Watson 1968, p. 10). Crick was a physicist who had changed directions in his research and developed a love of molecular biology. According to Watson, Crick terrorised his English colleagues, who feared that he might 'expose to the world the fuzziness of minds hidden from direct view by considerate, well-spoken manners' (Watson 1968, p. 10). The custom in England during most of the 20th century was to respect another scientist's territory and in so doing respect the scientist. This meant ignoring an interesting problem if it turned out that another scientist was already working on it. The rationale for this rule was simple enough: unless researchers could have some assurance of receiving credit for their discoveries, they would otherwise not devote enough time and effort toward making the discovery in the first place. Each scientist claims dominion over an area of inquiry, and in this way the attribution of authorship to the discovery would be a simple matter. At this time, according to Watson (1968, p. 15), 'molecular work on DNA in England was, for all practical purposes, the personal property of Maurice Wilkins'.

Wilkins at the University of London had the remarkable X-ray diffraction photographs but Crick was far enough away in Cambridge not to think too much about rationalising the results of the images. The custom of a separate and distinct division of labour was about to be breached. Watson was now at Cambridge and he would broker information between London and Cambridge. The Crick-Watson hypothesis – that DNA is shaped like a double helix – might not have occurred when it did had Crick and Watson adhered to the then existing unwritten rules of British professionalism. Rather than allow islands of separate and non-communicating research to persist, Crick and Watson formed an impromptu research team to broker the flow of information between the two academic ports. This required flaunting the rules, customs and other institutions that surrounded scientific research and protocol at that time. This middleman activity proved to be important to the final breakthrough discovery.

There were further complications as well. In Wilkins's laboratory at the University of London another scientist, Rosalind Franklin, was using X-rays to actually photograph DNA. It was obvious to Crick and Watson that Franklin jealously guarded her findings and would not willingly share her experimental data with them. Her jealous behaviour was the custom under the institutions of British science. Perhaps Franklin was hoarding her findings so she could publish the results first and obtain the most professional credit. In any case, it is clear that this rule of refusing to share results may have functioned to promote discovery when research was more personal and individualised. The Cavendish team did get her data but indirectly, without her knowledge or permission.

Apparently, Franklin had reported her data and images to the British Medical Research Council, which was looking into the activities of the laboratory in which she worked. Had she known that the information in that report would be reviewed by her rivals in Cambridge, she would have been both shocked and angry. Reviewing her research made clear that, whatever the result of the Crick-Watson collaboration to explain the structure of DNA, it would have to accord with Franklin's measurements. It was convenient for Crick and Watson to have these data when they did.

Franklin and other British scientists of that day might have dismissed Crick and Watson as eccentric quacks, for one important reason: they were willing to think about the structure of DNA by first building actual three-dimensional models. The models looked like children's toys. Apparently, it was not the custom in the 1950s to build models as a method of heuristic illumination. Crick and Watson used the models to help them 'visualise' both the overall shape and the connections within the shape of the DNA molecule. This technique was copied from the work of the celebrated American scientist Linus Pauling, but the unorthodox nature of the exercise indeed raised eyebrows in England. Again, the Crick-Watson team was breaking with tradition and revising the rules of good professional behaviour as well.

There came a point in early 1953 when Watson and Crick learned that Linus Pauling had proposed a model for the structure of DNA that was precariously close to their own thinking, but erroneous. Watson and Crick knew about the error that Pauling had made because they had made a similar error only weeks earlier. Should they quickly inform him of the problem or let him fall on his face by his public announcement of what they knew to be a false solution? The Watson-Crick team chose silence. Pauling would subsequently publicly admit his false start.

The fact that Pauling was so close to them in his research suggested to Watson and Crick that they had only six or so weeks to solve the problem before Pauling beat them to it. As Watson recalled, they met that deadline but not without an effort to keep the last stages of their thinking a secret and away from Pauling. Now the Watson-Crick team adopted the secrecy norm for their work, although they had previously denied the same courtesy to their rivals. The rivalry and the questionable methods used resulted in a decisive win for the Watson-Crick team and their model of the double helix.

Pauling found out about their discovery from Max Delbrueck, a scientist in Pauling's laboratory who corresponded with Watson. Watson was corresponding with Delbrueck in order to keep tabs on Pauling and how close he was to the solution to the DNA problem. Finally, in a letter dated 12 March 1953, Watson explained the solution to Delbrueck as a ploy to nail down priority for the discovery, but cautioned Delbrueck not to tell Pauling until the notice of the discovery had been sent to the journal *Nature*. Delbrueck could not resist telling Pauling and did so shortly after receiving the letter. Why did he 'spill the beans'? According to Watson, 'Delbrueck hated any form of secrecy in scientific matters and did not want to keep Pauling in suspense any longer' (Watson 1968, p. 217). But it was secrecy and the retardation of scientific information that the Watson-Crick team needed so that they could claim priority in the discovery and claim the Nobel Prize. We have seen that the Watson-Crick team did not respect this norm for other scientists, as when they obtained access to Franklin's measurements without her knowledge or permission. They chose to selectively enforce the old background institutions of scientific research. Some might characterise this behaviour as 'despicable'.

Watson's account ends with their discovery and the eventual happy reaction of all the scientists, including the always gracious Pauling. When the race had ended, genuine congratulations to the victors were in order and all forms of subterfuge and stratagem could again take a back seat. With the autobiographical account of the young Watson, we learn (1) what many customs and norms of the scientific community at mid-century were and (2) how Watson's team was to secure a victory in a rivalry that ended at the doorstep of the Nobel Committee in Stockholm. By carefully studying

Watson's own personal account of this episode in the history of science we obtain knowledge about the broader institutions that typically guided research among and between scientists. From this we can speculate about how these institutions normally functioned to promote an end that may not have been part of the original intention of those who behaved in accordance with these institutions. We also learn how two researchers, Watson and Crick, could advance their own progress by flaunting the rules, habits and customs of the scientific community while at the same time relying on others to adhere to those same rules. This episode highlights materials and patterns of behaviour that would have interested Menger and the later Austrians, since they were spontaneously formed but may no longer have functioned to produce beneficial results in the age of team research.

## 6. CONCLUSION

And so there you have it. I have offered a suggestion about how a genuine economics of science might proceed with historical information that is already at hand. I do not know if what I offer here meets Sent's criteria for a more 'synergetic approach to "science" and the "economy"' since Sent did not make it clear what such an approach might look like (Sent 1999: 115). I do believe, however, that my approach takes its starting point from an important topic of interest among modern Austrian School economists and relates it to the study of science and its organisation. Since the Austrian School emphasis is on institutions – by which they mean existing habits, customs and rules of behaviour – the tasks of an economics of science are to (1) identify as precisely as possible these habits, customs and rules of behaviour; and (2) assess the extent to which they function to produce certain results. In the two case studies I summarise above, the rules that promoted team-based research were either totally misunderstood (as in the Banting-Macleod debacle) or required that for one team to get to the finish line, older customs and norms of behaviour that guided the scientific community had to be selectively applied (as in the Watson-Crick episode).

I insist that it is possible and desirable to list those things that have emerged that help make the system of scientific communication, organisation and development more effective. Some rules, habits and customs no doubt have emerged spontaneously or organically. Other have come into existence through human intelligence and artifice quite suddenly, and partly in response to the emergence of the modern research laboratory and the imperatives of team-based research. How those rules have functioned to select scientists, authenticate experimental results, award prizes and corroborate facts both in general and during specific episodes of rapid change in understanding of matters scientific seems to me to be a terrific research project. In the short space of this chapter I have only been able to sketch what such an approach can look like. I have tried to further one theme suggested by the editors of this volume, namely, the connection between science and spontaneous order.

## NOTES

1 Also, rules that originated out of custom and habit can be a fetter on progress however defined since the origin of a rule and an appreciation of its functioning are two entirely different lines of investigation and analysis.

2 According to the American philosopher John Searle, status relationships – those that allow us to know so much and accomplish even more – are created by way of 'speech acts' such as the locution 'I now pronounce you husband and wife'. Not everyone can pronounce a couple 'husband and wife', only certain people who are recognised by the political jurisdiction within which the couple plans to reside. Consider those who work as 'justices of the peace'. How did they obtain that status? Clearly, there is some authenticating board that applies a set of standards to decide who in the community shall have the 'justice of the peace' status. Even more sensational, not all willing couples are eligible for the husband and wife status (for example, certain legal jurisdictions prohibit homosexuals from obtaining this status relationship). Where did all these complicated and time-tested institutions come from? Did they all arise spontaneously as a result of human action but not human intention and calculation?

3 Before turning our attention to the progress of science section of this chapter, I wish to clarify an important point. One modern 20th-century Austrian School writer – and a most extraordinary one at that – Hayek, argued that spontaneously generated rules that individuals adhered to unconsciously were often of greater importance and value to social stability than other rules and customs such as those arrived at by conscious planning and deliberation. On this point Hayek seemed to have been alone within the larger Mengerian Austrian School tradition. Contrary to Hayek, I shall adhere to the sharp distinction between the problem of the origins of the rules and norms versus their continued value and justification. As a matter of logic, the two sets of questions should not be confused. Also, the fact that the persistence of a rule contributes to the maintenance of a social system does not explain the continued existence of the rule itself. Hayek must specify the 'process by which a rule that is advantageous to the system is sustained in operation within that system' and the mechanisms by which it is passed on from one generation to the next (Hodgson 1993, p. 168). In the short compass of this chapter I must pass over some of these issues, but at least in this note I want to emphasise their importance to any economics of science.

## REFERENCES

Arthur, Brian W. (1994), *Increasing Returns and Path Dependency in the Economy*, Ann Arbor, MI: University of Michigan Press.

Barry, Norman (1982), 'The Tradition of Spontaneous Order', *Literature of Liberty*, **5** (Summer): 7–58.

Bliss, Michael (1982), *The Discovery of Insulin*, Chicago, IL: University of Chicago Press.

Hayek, Friedrich A. von ([1988] 1989), *The Fatal Conceit: The Errors of Socialism (Collected Works of F. A. Hayek, Vol. 1)*, William W. Bartley (ed.), Second edition, Chicago, IL: University of Chicago Press.

Hodgson, Geoffrey M. (1993) *Economics and Institutions: Bringing Life Back into Economics*, Cambridge: Polity Press.

Kuhn, Thomas S. (1977), *The Essential Tension: Selected Studies in Scientific Tradition and Change*, Chicago, IL: University of Chicago Press.

Menger, Carl [1883] (1963), *Problems of Economics and Sociology*, L. Schneider (ed.), Urbana, IL: University of Illinois Press.

Mowery, David C. and Nathan Rosenberg (1998), *Paths of Innovation: Technological Change in 20th-Century America*, New York, NY: Cambridge University Press.

Sabooglu, M. (1996), 'Hayek and Spontaneous Orders', *Journal of the History of Economic Thought*, **18** (Fall): 347–64.

Searle, John R. (1995), *The Construction of Social Reality*, New York, NY: Free Press.

Sent, E.-M. (1999), 'Economics of Science: Survey and Suggestions', *Journal of Economic Methodology*, **6**: 95–124.

von Mises, Ludwig ([1949] 1963), *Human Action: A Treatise on Economics*, New Haven, CT: Yale University Press.

Watson, James D. (1968), *The Double Helix: A Personal Account of the Discovery of the Structure of DNA*, New York, NY: Atheneum.

# 7. An evolutionary approach to the constitutional theory of the firm

**Jukka Kaisla**

## 1. INTRODUCTION

This chapter argues that contractarian reasoning can contribute to new institutional theories of the firm by emphasising the procedural justification of the constraints within which actors make choices among alternatives. Normative individualism as the methodology of constitutional economics brings with it logical constraints on efficiency considerations that new institutional theories are largely silent about. A central difference between constitutional economics and the new institutional theories of the firm is that while the latter focus mainly on efficient *outcomes* within a given institutional framework, the former examines the criteria by which the institutional framework *itself* can be considered efficient.

The chapter will maintain that taking *voluntary exchange* as the ultimate source of justification in constitutional economics is problematic. This is because the boundary between voluntariness and coercion depends on conventions of property and fairness which define how voluntary exchange is arrived at in the first place. I will argue that limiting efficiency considerations to voluntary exchange alone breaks with the constitutional procedural justification, producing logical inconsistency. I will propose that cutting the infinite regress of the procedural justification of rules should not be done before conventions have been incorporated into the justification process.

The firm can be seen to be constituted by a group of self-interested people cooperating and competing within a set of multi-layered rules. Individual decisions and actions are interrelated and coordinated in ways that allow us to refer to *corporate* (Coleman 1990) or *concerted* (Vanberg 1992) action. The contribution of the constitutional approach is that it highlights the (explicit or implicit) constitutional agreement as an *exchange of commitments*

(Vanberg 1994, p. 140). The contracting parties benefit from constraining their future choices within the constitutional framework. The core argument of the constitutional theory of the firm is that an organisational social contract results in relations among the parties that are different in kind from market relations (see an opposing view in Alchian and Demsetz 1972).

There are not many theorists who have analysed the firm from the contractarian perspective. Vanberg (1992) has provided a persuasive analysis of how the constitutional paradigm can provide a consistent individualist interpretation of organisations as acting units. His approach is linked with Coleman's (1990) analytical perspective on the procedural foundations of collective action. Gifford's (1991) constitutional analysis of the firm argues that the firm will benefit if relation-specific investments can be secured through the owner's attempt to purposefully design an efficient constitution. Wolff (1997) recognises that corporate culture, as in Kreps (1990), can be taken as an implicit part of the constitution of a firm. Langlois (1995) discusses the interplay between constructed and spontaneous elements in the emergence and perseverance of firms.

Figure 7.1 illustrates the goal of this chapter. The contributions of Coleman, Vanberg and Gifford are depicted as area A, providing a constitutional approach to the firm. My aim is to justify the introduction of conventions into the contractarian perspective (area B), and to propose that taking into account the interplay between evolution and design can aid our understanding of many aspects of the firm (area C).

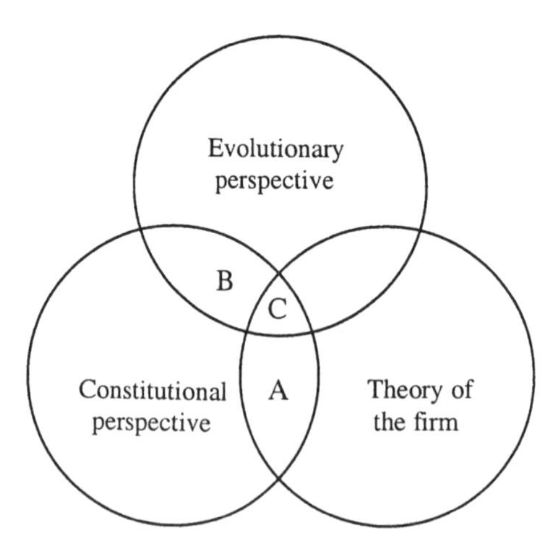


*Figure 7.1 Connecting themes*

The extended constitutional approach may have some interesting connections with other theories of the firm. I propose that it can be viewed as an overarching perspective that can encompass many existing insights into the theory of the firm. It can provide further explanations of the resolution of *coordination* and *motivation* problems within organisations, emphasised by modern contract theory (cf. Milgrom and Roberts 1992). Matters such as the feasibility of incomplete contracting, or why participants are willing to constrain their rent-seeking and opportunistic tendencies in relations characterised by asset specificity and asymmetric information, receive explanations that complement transaction cost and other contract theories. Corporate culture has a close connection to the present theme and will be discussed in connection with constitutional reasoning.

The chapter is organised as follows. Section 2 will provide some basic principles of constitutional economics. It will also initiate the rationale for directing constitutional analysis to the theory of the firm. Section 3 aims to explain the rationale for introducing conventions into contractarian reasoning. In Section 4 I turn to examine the constitutional theory of the firm. The discussion is limited to the contributions that are closely related to the contractarian reasoning provided by Buchanan and other advocates of constitutional political economy. The reason for this limitation is based on the recognition that the term 'constitution' may be used in various ways that do not have to correspond with contractarian philosophy. In Section 5 I will examine the extended constitutional perspective further in relation to other theories of the firm. Finally, Section 6 will provide some concluding remarks.

## 2. ON THE BASIC PRINCIPLES OF CONSTITUTIONAL ECONOMICS

Constitutional economics is essentially about the examination and evaluation of the foundational rules of social order. It is an inquiry into the interrelation between what Hayek called the *order of rules* and the *order of actions* (Hayek 1973). The constitutional perspective suggests that in our pursuit of social improvement, changes in the order of rules ought to be the principal means (Vanberg 1994, p. 5). It directs our analytical attention toward the *choice among constraints* (Buchanan 1991, p. 5). This perspective implies the recognition that societies are complex systems where purposeful design directed to particular outcomes does not in many cases bring about what is desired. This results from the genuine uncertainty of outcomes. The source of uncertainty is our ignorance of the unintended consequences inherent in human (inter)action. Although an organisation can be viewed as being intentionally constructed to realise a certain purpose, the actions taken within the organisation have unintended consequences as well. This relates to Hayek's view on the relation between the origin of rules and their outcomes as he states that 'it is possible that an order which would still have to be described as spontaneous rests on rules which are entirely the result of deliberate design' (Hayek 1973, p. 46). What is meant by the notion of *purpose* becomes central. If 'purpose' refers to rather concrete ends, it is consistent to view the goal of an organisation as an outcome of purposeful design. This, however, leaves open the question to what extent the attainment of that goal can be viewed as a planned process. Alternatively, if one views purpose as being directed towards more abstract ends such as self-maintenance or survival (cf. Selznick 1948), the purpose seems to owe more to spontaneous elements.

The constitutional perspective directs the analytical interest away from the goal-oriented discussion to the foundations of agreement on participatory and distributional rules. The firm is then not defined through its possible goals, but through the rules that constitute a system of productive relations among the participants. The constitution of an organisation specifies the terms of participation: (1) which resources participants are to contribute to the organisation; (2) how and by whom the decisions on the use of pooled resources are to be made; and (3) how the resulting benefits from the joint endeavour are to be shared among participants (Vanberg 1985, p. 22).

Constitutional analysis is consistently individualistic. (1) The derivation of institutional constraints is based on a calculus of individual interests. (2) Collective choice is derived from the participatory behaviour of individual members. (3) Emphasis is directed to the selection of rules that will limit the behaviour of those who operate within them (Buchanan 1991, p. 8).

The methodology of *normative individualism* provides a normative point of departure for the constitutional economics advocated by Buchanan and other contractarians. Normative individualism suggests that we should take the values and interests of the individuals involved as the relevant standard against which the goodness of rules and their outcomes is to be judged (Vanberg 1994, p. 1).

The constitutional perspective highlights voluntary exchange as the core motivator for the individual to limit his or her behaviour within constraints. The cost of limiting one's own behaviour is accepted insofar as it does not exceed the benefit resulting from reciprocal behaviour of others. This perspective emphasises the calculative rationality of the individual who actively chooses his or her own constraints. By definition, a voluntary exchange happens only when the participants expect to gain from the trade.

The subjectivist position of the constitutional perspective recognises that values and theories about the world vary across individuals. This limits efficiency considerations because it is believed that no supraindividual scale of goodness exists. There is no reason to believe that the ordering of preferences would not vary in time and across individuals.

From the subjectivist position, an assessment of efficiency relies on revealed preferences of the individual. When the idea of voluntary exchange is transferred to the realm of collective choice, the strict criterion of revealed preferences through observed exchange needs to encompass all the parties. As the subjectivist position holds that the values of individuals are incommensurable, an exclusion of any one party from the exchange breaks down the possibility of verifying that the observed exchange was in fact efficient.

Since individuals vary in their tastes and interests, it is likely that when a group of people get together in order to pursue something collectively, conflicts of interests arise and mutual agreement may thus be difficult to achieve. The members need to *compromise* before a mutually agreed solution can be reached. The solution may not match perfectly with anybody's immediate interest but provides a more desirable outcome than being without it. The question about how to facilitate a compromise thus becomes central. A compromise requires the parties, to some extent, to alienate their immediate self-interests and, through introspection, assess what would be considered fair by the other parties.

The constitutional bargaining process itself contains aspects that facilitate agreement (Brennan and Buchanan 1985, p. 29). Rules are by definition more general than the outcomes that result from action guided by those rules. A constitutional choice among alternative rules contains the elements of generality as a chosen rule needs to be applicable in numerous contingencies. Another basic characteristic of a rule is its extended time dimension. A rule needs to be applied over time, otherwise it can hardly be considered a rule. Due to these considerations, the individual faces genuine uncertainty about how his or her position will be affected by the operation of a particular rule. Insofar as mutual agreement is the goal, the individual tends to agree on arrangements that can be considered fair in the sense that they are broadly acceptable (Brennan and Buchanan 1985, p. 30).

For a contractarian, the only justified criterion of goodness in collective choices is a unanimous agreement among the participants. Alternative constitutional arrangements can be analysed in a hypothetical initial state to discover what basic principles such rules should fulfil. This may help him or her to create new alternatives that may receive acceptance among the relevant group. But, as Buchanan has recurrently noted, the members of the group are the sovereign decision-makers whose individual values are the only justified source for efficiency considerations. What makes this central principle problematic is its consequence for innovative and creative aspects of social endeavour. It is obvious that many innovations are such that only few understand their potential value immediately. Changes in rules are especially difficult to negotiate because individuals generally value the status quo (Schlicht 1998). Another problem that Barry (1984) has pointed out arises as all the members are in a position to veto an alternative that would otherwise be desirable. From the contractarian position, there is of course no such thing as 'otherwise desirable', but it is intuitive to think that somebody may want to veto whatever the rest of the group are suggesting. This connects to the pragmatic criteria of justifiable exclusion from decision-making (children, mentally challenged persons and so on). Any agreement on the contents of such a list fails by necessity to meet the contractarian ideal, however. This is because the exclusion must occur before the list is agreed upon.

Contractarians have tried to resolve this problem by proposing that not all choices among rules need to satisfy the strict criterion of unanimity. It is entirely justifiable for any group to unanimously agree upon relaxing the criterion for post-constitutional rules that are specific to the extent that a complete agreement would be too costly to achieve (Buchanan and Tullock 1962). This makes the perspective more operational but also creates a logical problem. If post-constitutional rules may be placed in categories of various degrees of unanimity (simple majority, two-thirds, three-fifths, and so forth), then a choice of which category to use regarding a particular post-constitutional rule becomes central. If a choice about the proper category of a rule is *less* than unanimous, then the choice *itself* becomes unjustified on constitutional grounds. The participants know that the higher the degree of unanimity that is required, the less probable it is that a rule will become accepted. Thus, those who favour a certain post-constitutional rule try to get it into a category with a low degree of unanimity, whereas those who oppose it try to get it into the category of the highest level of unanimity. Therefore, if such a choice itself is not unanimous, there is no guarantee that any post-constitutional rule is efficient.

One can attempt to remedy these problems concerning the strict unanimity criterion in three ways. The first alternative is to relax the strict unanimity requirement and accept a more operational alternative that takes unanimity as an ideal. The goodness of a collective choice is then measured by its *degree* of unanimity. The problem with this alternative is that it opens the door for the Kaldor-Hicks-type reasoning by which a change in rules would be acceptable insofar as the benefit would over-compensate the loss and those who win could hypothetically compensate those who lose. The contractarian position maintains that in such a situation the compensation should be observed, otherwise no guarantee of mutual benefit is provided. This argument is logically problematic, however. If the participants perceive and calculate the compensation, there is no reason for them *not* to agree. The result then would be unanimous agreement. If, on the other hand, the participants do not perceive the compensation as a part of their gain in exchange (perhaps due to ignorance), the normative individualistic principle of the contractarian position does not allow an agency to 'know better' what is good for the participants, and thus the change in rules would remain coercive. To be sure, the idea that a change in rules is justifiable in contractarian terms insofar as losers are compensated is unhelpful as it directs attention to *outcomes* that are nonexistent at the moment of choice. After all, in contractarian reasoning, efficiency of a collective choice is not derived from *ex post* outcomes but, instead, from the exchange at the moment of collective choice.

The second alternative for remedying problems arising from the strict criterion of unanimity is to limit the applicability of contractarian reasoning to rules that are general enough to facilitate agreement behind the veil of ignorance (Rawls 1971) or of uncertainty (Brennan and Buchanan 1985). It is likely that general *principles* of human rights, separate property, and government may be shared in a large group. The problem with general principles is that in order for them to carry behavioural influence, they need to be *interpreted* in dissimilar situations. Even though an agreement on general principles may be reached, unanimity on rules that define more accurately the scopes and limits of various (often conflicting) rights may remain unattainable, as the participants are able to foresee how a certain configuration will affect their relative positions.

The third alternative for remedying problems arising from the unanimity criterion is to direct contractarian reasoning to smaller groups whose members may share more coherent sets of values. This idea is based on the assumption that the more coherent the values among the participants are, the more detailed rules the participants can agree upon. Another central assumption is that in smaller groups the probability of more coherent values is higher than in larger groups. If these assumptions prove to be realistic, contractarian reasoning is perhaps more applicable in smaller organisations, such as business firms, than at the level of nation states.

The rest of this chapter aims to examine this third alternative. However, even though agreement on more detailed rules may generally be better facilitated in smaller groups, I shall maintain that the exchange of commitments among the participants depends on the conventions of fairness and reciprocity in that group. This view is closely related to Sugden's interpretation of contractarianism in which 'the object is to evaluate possible changes in the institutions of an existing society, using a criterion of agreement that is defined relative to the knowledge and the *conventions* that prevail in that society' (Sugden 1993: 421, emphasis added).

## 3. ENTER CONVENTIONS

In the approach proposed here, conventions enter the constitutional analysis in two distinguishable ways. First, it is maintained that from a structural perspective conventions logically precede a social contract. Second, it is argued that a social contract upon a rule does not *per se* produce coordination of actions and expectations. A rule needs to be continuously reinterpreted as the future is disclosed. It is the ongoing interpretative effort that produces shared expectations of appropriate courses of action.

The term 'convention' is here broadly defined as a social pattern of behaviour which is shared among the participants to the degree that it is common knowledge that most participants conform to the pattern and that most participants expect most of the others to conform as well. Conventions can be contractual and noncontractual. A noncontractual convention is one which does not require a covenant to enforce conformity because it is in the direct interests of everyone to conform; and there is no need for explicit agreement as it is obvious to all parties what the expectations of others are. A convention, such as on which side of the road to drive, is thus noncontractual in the sense that no contract needs to be established for its maintenance (Gauthier 1998). A convention is contractual if its maintenance requires an agreement and a covenant to enforce the agreement. For instance, a rule by which a firm's outcome is to be distributed among the participants is contractual.

As a first approximation, the term 'contractual convention' might be used interchangeably with the term 'social contract'. They both require agreement. The distinction between a social contract and a convention can be seen in their dissimilar relation to time and process. The constitutional position emphasises a two-step notion of rules: before a game can begin, rules must be agreed upon. In the second step, the game is carried out within the agreed rules (Buchanan 1991). This leaves open the question about how those rules are to be interpreted in disclosing future situations that are necessarily nonexistent at the moment of rule-making, and also, how interpretation modifies the rules themselves. A social contract can, of course, be seen as conjectural in the sense that the disclosing future can be expected to provide feedback so as to motivate the participants to revise the contract to better correspond with changed circumstances (Vanberg 1986). However, the way participants come to recognise a decrease in the desirability of the present rule is argued here to be due to a recognised discrepancy between the *de facto* convention that is adhered to and the rule that was agreed upon in the past. It is unrealistic to assume that the participants collectively and simultaneously suddenly realise the presence of a discrepancy that they had all failed to recognise earlier.

Thus, it is argued here, it is the shared interpretation of a rule that influences behaviour, not the agreed rule itself. Such a collective interpretative effort has more to do with convention-formation and change than with the rule as an outcome of collective choice. Even though this argument may sound somewhat strange, consider the following logic. The explicitly agreed rule itself is a *principle* that has been arrived at through collective interpretation. I argue that the interpretative process by which the principle is arrived at is more important than the principle itself. This is because what explains the agreement upon a rule is not the rule itself, but the collective interpretative effort that gives rise to agreement. Another reason for this argument is not derived from the process by which an agreement is arrived at, but from the process of interpretation that begins immediately *after* the rule is in place. After an agreement upon a rule, interpretation of its meaning begins to have its influence. Viewed this way the explicitly agreed rule itself serves as a discontinuation point in the *processes* that both create it and interpret its meaning.

The first argument for the structural priority of conventions over a social contract, as presented in the beginning of this section, is derived from the logical problem that would result if no convention were assumed to exist at the moment of agreement. As Block and DiLorenzo (2000: 571) put it:

> Constitutional economists try to derive a theory of human and property rights from their constitutional framework and they seek to do so on a consensual basis. But how can people give their consent to a contract before it is clear that they have any rights to do so? Where do these rights come from? How can a person agree to be bound by a constitution if it is this very document which can alone establish his rights? If rights are established only by constitutions, then before their advent individuals have no rights. But if they have no rights, what 'right' do they have to participate in the construction of a constitution?

Taking an agreement as the starting point is logically problematic because if the initial state does not already include some mutual expectations, that is, a convention of reciprocity, a social contract remains unattainable. This refers to Hobbes's ([1654]1996) model of the initial state. In that model, the participants cannot resolve the first-mover paradox because in a genuine anarchy, reputation accumulation is impossible. In order for a protective agency to arise, the model must be extended to include at least bilateral reciprocity in a way presented by Nozick (1974). The reason why I use the term 'bilateral reciprocity' rather than 'exchange' is that although there is exchange, it is not temporally symmetric in the sense that all exchanging parties receive gains at the moment of exchange. They receive expectations of gains that require the keeping of promises and trust.

The contractarian position is maintained so as to be able to systematically extend the individualistic perspective of classical liberalism into the realm of collective choice (Vanberg 1994, p. 204). The individualistic position maintains that voluntary exchange indicates agreement among the parties, and that such voluntary agreement is the ultimate criterion on which an exchange can be judged to be efficient (Buchanan 1977, p. 128). By direct analogy, the contractarian individualistic position maintains that a collective choice can only be judged efficient if it is based on voluntary agreement by all parties involved (Vanberg 1994, p. 204). This chapter maintains that the way voluntariness is defined in particular situations depends on the relevant conventions among the participants.

## 4. THE CONSTITUTIONAL THEORY OF THE FIRM

Firms, like other organisations (clubs, associations, states, and so on), are constituted by their members. By entering into an organisation a member becomes subject to the authority system of that organisation. An individual voluntarily gives up some of his or her autonomy in return for the benefit he or she gains from participation. When entering an organisation the individual not only accepts the authority system, but is also willing to submit part of his or her resources to be pooled and subjected to unitary control. It is through the exercise of control over the pooled resources that an organisation can meaningfully be treated as an acting unit. The constitution of a business firm states the terms of membership as well as the member's rights of participation in controlling the combined resources (Vanberg and Buchanan 1986: 216). Many desirable aspects in the firm dynamics depend on the success of coordinating efforts among the members and on the ways that rights are defined and justified. Capability accumulation, knowledge creation and dissemination, communication and coordination of plans and actions are examples of such aspects. It seems reasonable that we should direct our analytical interest toward the constitutional dynamics of firms when long-term developmental issues are studied.

The constitutional rules of an organisation can be described as solving two types of problem: those arising (1) in team use of pooled resources and those arising (2) when the social product of collective endeavour is distributed among the members (Vanberg 1994, p. 139). The former type of problem refers to *knowledge* problems of how to arrange and coordinate various tasks within the organisation. The latter type of problem seems to correspond better to the *conflictual* aspects of self-interested members. The central criterion for agreement is that a rule needs to be general enough to facilitate impartial judgement. Rules of distribution do not necessarily provide uncertainty to the extent that the members could not foresee how their positions would be affected. With regard to privately-owned business firms, an equal-share rule is not more prominent than any other alternative. One solution to alleviate conflicts of interest in distribution is to examine how far property rights can be developed to provide prominent demarcation in collective endeavour.

It appears intuitively obvious that if the property rights within a firm are ill-defined or ill-protected, the members suffer through reduced incentives to put in effort and increased incentives to rent-seeking. Gifford is in line with other asset-specificity theorists when he recognises that the core problem arises when it would be in the interest of the firm to have the members making firm-specific investments (Gifford 1991: 91). If property rights are ill-protected, members remain vulnerable to rent-seeking on the part of others and thus remain reluctant to make such investments. The constitution of a firm is then viewed as a remedy for this undesirable state of affairs. The constitution is seen as a set of interdependent long-term contracts among the members (Gifford 1991: 92).

The role of the constitution, for Gifford, is to 'set up a system of constraints, limiting the ability of individuals and coalitions to impose external costs on others' (1991: 92). A constitution is thus designed primarily for limiting opportunism within organisations. For Gifford, the remedy for rent-seeking tendencies is a constitution created by the owner (or his or her agent) 'to maximise the sum of the present values of all the assets used in the firm' (ibid.: 93). The central purpose of the owner setting up constitutional constraints is to provide incentives for the employees to make firm-specific investments. This is accomplished by protecting the property rights of the employees to their firm-specific investments. 'By creating an efficient constitution the owner of the firm maximises the value of his own assets in the firm and at the same time those of the other firm members' (ibid.). The positive externalities that the owner thus creates are internalised by other firm members. This can partly explain the motivation for an individual to join a firm. A member can gain access to the pool of knowledge and is at the same time protected by the constitutional rules against potential rent-seeking by other members.

Furubotn (1988) is in the same line of reasoning as Gifford. By *codetermination*, Furubotn means a provision of control rights that give those employees who make firm-specific investments part of the firm's control rights. The decisive criterion is whether or not representatives of labour take part in the firm's decision-making processes at board level (Furubotn 1988: 166). The core idea is to explain that the firm maximises its profits by giving those employees who make firm-specific investments their share of decision-making rights. Furubotn maintains that the firm is actually a 'joint investment' among capital and labour providers and therefore the employee-investors should be regarded as equity holders (ibid.: 168). The sharing of control rights via codetermination is then maintained so as to provide some assurance that '*all* interests will be considered in decision-making and that unfair allocation of quasi rents will be prevented' (ibid.: 168–9, emphasis in original).

Furubotn's analysis remains noncontractarian in its emphasis on the profit-maximisation rationale for joint decision-making. To be sure, if profit-maximisation is the rationale for joint decision-making, then we have to consider the trade-off between increased decision-making costs on the one side, and the increased coherence between the rules and the interests of the participants on the other side (Buchanan and Tullock 1962). Such an assessment is necessarily directed towards consequences that are nonexistent at the moment of choice, and therefore remains speculative.

Gifford's analysis recognises the central rationale for a constitution as constraining the self-interested behaviour of the firm members. However, his analysis remains somewhat distant from the normative individualist foundation of constitutional economics. The idea that a central agent should (be able to) design an efficient constitution for the members of the firm to follow seems to disregard core issues in the procedural justification of contractarian reasoning – as well as the epistemic limitations of human actors. The unanimity criterion in constitutional economics is established precisely because of the problem that we cannot know whether a collective choice is efficient or not by any other means than by assessing the degree to which it corresponds with the interests of the relevant parties.

## 5. CONNECTIONS TO OTHER THEORIES OF THE FIRM

A common denominator for theories of the firm is that they are characterised by their concern with the existence, the boundaries and the internal organisation of the firm. Another common theme is that explanations for these matters are based on outcome-oriented efficiency considerations. The goal of the present approach is different. It discusses some foundational principles of a constitutional order within the business firm. The constitutional approach advocated here corresponds with the principles of subjectivism which give limited scope to derive efficiency claims. This chapter is thus unable to assess to what extent constitutional rules of an economic organisation are efficient in some other sense than being desirable, as judged by the members themselves.

The literature on the theory of the firm is expanding and it would be futile to try to discuss all the various approaches in this context, especially in a way that would throw any more light on the matter than has already been done by others (for a detailed discussion of various contributions see Foss 1999). In this section, I will discuss some ideas from different approaches that are connected with the main theme of the chapter.

The issues connected with the constitutional perspective that are of interest here concern the contractual arrangements within the firm. Interesting issues arise from coordination problems as well as incentive-conflicts among the members. Transaction-cost considerations correspond with our immediate intuition as well. An economising individual will prefer more goods to less and less 'bads' to more. It is therefore expected that people will try to organise production in ways that minimise various types of costs that necessarily arise from action. It is another thing to what extent lists of different kinds of costs take into account all relevant costs, or whether all those costs that influence choice behaviour can even in principle be made operational (cf. Buchanan 1969). Be that as it may, various approaches contribute to our understanding of the dynamics of economic organisations, a subject which is continuously changing as new, hitherto unperceived organisational arrangements are being created.

A constitution of a nation state applies to every member of that nation, even the legal-political elite (this is at least the general ideal of it). Things are not the same within business organisations, however. Power relations emerge not only through political processes within firms but are also part of the legal statuses of the members. Owners and managers have, in part, different sets of legal rights and obligations than employees. Therefore, constitutional considerations within economic organisations differ from those at the national level.

Coleman (1990, p. 327) has introduced useful terms that recognise these important distinctions; a constitution is *conjoint* when the beneficiaries and the targets are the same persons. A constitution of a Western nation is a good example of this. Although not every member of the nation participates to the same degree in the process of establishing the constitution, those who are targets, that is, those who are constrained by the constitutional rules, and those who benefit from having a constitution are the same persons. Every member faces both costs and benefits from constitutional constraints. Cost accrues as the individual has to constrain his or her own action within the limits of the shared rules. Beneficial impact comes from others' similarly constrained behaviour.

A constitution is *disjoint* when the beneficiaries and the targets are not the same persons. As an extreme, those who benefit from certain rules may be completely different individuals than those who are subjected to those rules. An owner-manager of a firm may design a set of rules that constrains the actions of his or her subordinates but which do not concern him or herself. It is sensible to argue that economic organisations represent constitutions that have more disjoint characteristics than can be found at the national level. As a first approximation, this could imply that business firms are characterised by more arbitrary rules than nations, and that subordinates within firms are subject to more coercive rules than their superiors.

## 5.1 Exit and Efficiency

Markets in which business firms are embedded provide prominent resolution mechanisms to the potential coercion of firms' constitutional arrangements. It is reasonable to argue that employees have better opportunities to vote with their feet, that is, to withdraw from a firm that enforces unjust rules, compared to emigrating from a nation state (Hirshman 1970; Wolff 1997). This fact alleviates the potential coercion within disjoint constitutions. Also, an employee's ongoing participation in a firm is taken as an *implicit consent* to the firm's constitution. Although exit is more operational when examining economic organisations than when discussing entire societies, it is not entirely unspeculative regarding economic organisations.

Exit indicates that the participant is not satisfied with the present constitutional order, or that a better alternative has been found. This may lead to a logical problem in the constitutional approach. Continued participation is assumed to reveal interest in accepting the organisational constitution. On the other hand, the normative individualistic foundation does not permit efficiency considerations other than those based on observed exchange. This means that when we observe two consecutive exchanges by the same actor, we cannot assess their comparative goodness based on procedural justification. This is because an observed exchange does not contain information about its relative efficiency regarding other observed exchanges (cf. Buchanan 1969, Lachmann 1976). What this implies in the context where a member of a firm decides to change their employer is that both having stayed in the present company and entering into the new one enjoy equal procedural efficiency. An attempt to argue that the exit is due to an unsatisfactory constitutional order is inconsistent with the procedural criterion of goodness of constitutional economics.

Constitutional economists probably accept the idea that each agreement is conjectural in the sense that it may become changed as circumstances call for it. Although one can argue that the change occurs because the old rule has become inefficient in the sense that the members do not perceive it as advantageous any more, one cannot argue that one somehow knows the comparative efficiency of the new rule over the old rule *when it was chosen.* This issue is central to how we perceive the change in rules and thus cultural evolution. What I am arguing here is that, based on our limited reason and imperfect knowledge, insofar as two consecutive choices are based on voluntary exchange there is no secure way for us to measure their comparative efficiency. Consider two consecutive choices made by a single chooser; the first choice is about which car to buy and the second choice is over a range of shoes. We cannot claim to know which one of these choices was more efficient based solely on the observation of exchange. To be sure, the chooser does not know it either – in the consequential sense, that is. The beautiful car he or she has bought may turn out to be a catastrophe while the shoes he or she bought in the sales may serve him or her well for years to come.

Properly understood, the constitutional criterion of goodness is only concerned with the realisation of the members' interests *at the moment of choice*. The constitutional perspective does not pretend to have foresight into the degree of consistency between expectations and outcomes that eventually unfold. That is why entering a firm at $t_0$ point in time and entering into another at $t_1$ point in time deserve equal procedural efficiency. As soon as the agent enters the new firm at $t_1$ it may become clear that the previous firm was the better alternative. But to know this requires accumulation of knowledge that was not there before $t_1$.

## 5.2 Incomplete Contracting

Incomplete contracting theories break with the Arrow-Debreu assumption of complete contracting. It strikes one as being rather realistic to assume that individuals do not know all the future contingencies which may affect the carrying out of a contract of any complexity or time-span. Despite this, both the nexus of contract approach and the formal principal-agent theory are largely based on the assumption of complete contracting (Foss 1999).

*Coordination* is one of the themes around which the incomplete contracting approach rotates, beginning with Coase's (1937) seminal contribution. Wernerfelt (1997), for example, argues that the firm exists because of its advantage in minimising *communication costs* in intrafirm relations. Herbert Simon (1951) emphasises the distinction between the employment contract and the market contract. This perspective contradicts another contractual idea, developed by Alchian and Demsetz (1972), that intrafirm contracts cannot be distinguished from market contracts. Their analysis implies that the firm is reduced to a fictitious legal entity. The constitutional perspective is founded on the recognition that intrafirm relations are essentially different from market relations. They are different enough to make the concept of *concerted action* operational within the firm (Vanberg 1994, p. 135). It is precisely the cooperative team dynamics, which are not decomposable into bilateral agreements among the members, that make intrafirm relations different from market ones (see also Coleman 1990). Simon (1951) argues that the advantage of the employment relationship over the market contract lies in its *flexibility*. After the employee has submitted to the governance structure of the firm, his or her action can be adapted more fully to unforeseen future contingencies.

*Asset specificity* is another theme in incomplete contracting. Unlike the coordination approach, the asset-specificity perspective highlights the organisational implications of *ex post* opportunism when relation-specific investments are involved (Foss 1999, p. 25). Williamson (1971; 1991) and his followers extensively discuss the implications of *opportunism* combined with Simon's concept of *bounded rationality* for different types of economic organisation. This approach resonates with the constitutional perspective of Gifford (1991).

Contracts of any complexity or time-span remain imperfect. This is due to our ignorance about how future events will affect what is agreed upon. Despite this anomaly, the parties can agree as new events disclose that certain implicit terms are binding and thus help in mending the initial contract. In order for the implicit terms to be effective, the parties must share their meaning. Otherwise the agreement breaks down. In order to secure agreement the parties submit to conventions that bring coherence to their interpretations of implicit terms. This is to say that an underlying reason for a successful application of implicit terms and contracts can be found in conventions.

## 5.3 Spontaneous Elements

The constitutional perspective of the present chapter differs to some extent from that of contractarian philosophy as defined in Brennan and Buchanan (1985). The present approach takes into account not only explicit agreements among firm members but also conventions. The perspective is related to approaches that emphasise the plurality and complexity of the relations within organisations. For instance, Herbert Simon states that:

> To many persons, an organization is something that is drawn on charts or recorded in elaborate manuals of job descriptions. ... In this book, the term organization refers to the complex pattern of communication and relationships in a group of human beings. This pattern provides to each member of the group a ... set of stable and comprehensible expectations as to what the other members of the group are doing and how they will react to what he says and does. (Simon 1976, introduction to the third edition, as referred in to Baker, Gibbons and Murphy 1997)

A number of writers within related perspectives share the understanding that implicit contracts and spontaneous procedures are essential components of organisational dynamics (see for example Barnard 1938, Granovetter 1985, Simon, 1976). The present study shares Barnard's view that many of the rules and practices are organisation-specific:

> [Consider] the lines of organization, the governing policies, the rules and regulations, the patterns of behavior of a specific organization. Though much of this is recorded in writing in any organization and can be studied, much is 'unwritten law' and can chiefly be learned by intimate observation and experience. (Barnard 1976, p. xliv)

The present perspective is also related to Baker, Gibbons and Murphy's (1997, p. 23) analysis of implicit contracts. They emphasise the role of management in 'the articulation of unwritten rules and codes of conduct, the development and maintenance of a reputation for abiding by these rules, and the use of subjective assessments and informal adaptation to events in the implementation of these rules'. In this chapter the approach deviates from theirs in that the emphasis is on the role of conventions as constitutional constraints. The creation of implicit contracts is therefore not seen as being as 'conflict-laden' a process as Baker, Gibbons and Murphy suggest. Kreps's (1990) emphasis on the role of *corporate culture* gives some insight into these matters.

## 5.4 Corporate Culture

In his analysis of corporate culture, Kreps discusses the realm that should reasonably be related to spontaneous processes within organisations. In his terms, corporate culture consists of 'the interrelated principles' that the organisation applies and 'the means by which the principle is communicated' to say 'how things are done, and how they are meant to be done in the organization'. Because corporate culture is 'designed through time to meet unforeseen future contingencies as they arise, it will be the product of evolution inside the organization...' (Kreps 1990, pp. 93–4). Corporate culture does not only consist of the basic principles, but plays a role 'by establishing general principles that should be applied' (Kreps 1990, p. 126). This may be taken to be related to the evolutionary idea that once a convention has been established, it becomes a reference point for future development.

The reason why the employees of a firm have reason to expect authority to be used fairly is their expectation that *reputation* is considered a valuable asset (Kreps 1990, p. 92). I would suggest that reputation alone does not ensure fairness in adapting to unforeseen future contingencies. We need something that links reputation to new situations. That link is suggested to be in the form of conventions that provide shared interpretation of fairness and also a potential to establish shared reference-points for new events as they disclose themselves.

The approach in this chapter deviates from the analysis by Kreps (1990, p. 130) in that it does not assume any single and rigid focal principle. When discussing the optimal size of an organisation, Kreps (1990, p. 129) assumes that a corporate culture faces problems when the span of the principle is increased. This is because the range of contingencies that the principle must cover must also increase. The applicability of the principle (or culture or contract in Kreps's terminology) becomes ambiguous when increasingly dissimilar contingencies are introduced. A potential reason for this interpretation may be the disregard of rules in shaping interpretations of new contingencies. The essence of any rule is that it applies to a range of dissimilar events but what is equally important is that our perception of unexperienced events is based on our capacity to perceive them through *categories* of events, not as unique events as such (Hayek 1952). This alleviates the claim that when there is a gradual expansion of contingencies (organic growth of the firm) the rule necessarily becomes increasingly ambiguous. The relative rate of change between the categories of contingencies and the rule itself then becomes the key issue. External shocks aside, there is no *a priori* reason to assume that the change in a rule could not correspond with the changes in categories.

In Kreps's analysis, corporate culture seems to acquire a rather rigid interpretation. The situation is not alleviated by the use of interchangeable terms: focal principle, implicit contract and corporate culture (Kreps 1990, p. 130). If we assume only one focal principle or implicit contract applied in an organisation, there is reason to believe that, be it however clear and prominent, it does not provide much behavioural guidance in unforeseen future contingencies. But if we assume that there are several principles, and perhaps conventions, things change. For Kreps, this is not a solution, though, as he claims that a wider range of principles 'may increase ambiguity about how any single contingency should be handled' (ibid.). The reason for his doubt may be found in his general approach to corporate culture as being constructed by purposeful design. From the constructivist perspective the working properties of new principles are always uncertain and may only confuse the members of an organisation. My suggestion should at this point be rather obvious. I view corporate culture as being constituted by a system of conventions as well as designed principles. Conventions facilitate a wider range of principles without necessarily increasing ambiguity in interpreting unfolding contingencies. On the contrary, a central aspect of rules is that they shape our interpretations of dissimilar events. Even in the purest form of situational analysis, where we negotiate a situation which we have no previous experience of, we try to form a solution by referring to elements that bear some resemblance to our existing categories of recurrent patterns. This dynamic is often overlooked, resulting in an unwarranted picture of our choice processes as being distant from rule-following as a behavioural disposition.

In my terminology, corporate culture would be closer to the notion of the organisation's spontaneous order, which, although it partly results from rules and principles of designed origin, should not be taken as fully designed. In this reconstruction, conventions play a role as well as explicitly agreed rules in creating corporate culture.

Another difficulty arises in Kreps's analysis because of its static nature. Kreps (1990, p. 126) states that 'efficiency can be increased by monitoring adherence to the principle (culture). Violation of the culture generates direct negative externalities insofar as it weakens the organization's overall reputation'. In Kreps's treatment, corporate culture is (nearly) tangible. It seems to be easy to observe when it is strengthened, as well as when it is weakened. Both violations of the culture and their consequences seem to be readily measurable. Insofar as we remain in static analysis, corporate culture remains unaltered when all the parties follow it. Kreps claims that '[r]ewarding good outcomes that involve violations of the culture generates negative externalities [because it] weakens individual incentives to follow the principle and thus increases (potentially) the costs of monitoring and control'. The static perspective of his analysis makes changes in corporate culture unfeasible. Any experimental activity is *a priori* pronounced detrimental.

In this chapter's approach, experimental activity is central to the notion of change in human and social affairs. Although conventions may be unresponsive to situational variations, they will not remain unaltered. Even technological standards, which may, for a period of time, preclude alternative arrangements from emerging, will eventually give way to something new (see for example Constant 1980). In order for a convention to change spontaneously, somebody may initiate change by violating the existing convention. The violation does not have to be dramatic in the sense that it may still be based on some other convention, such as general reciprocity, and receive its justification from that. Also, existing conventions may promote the emergence of new alternatives.

## 6. CONCLUSION

The broad goal of this chapter has been to promote an understanding of the linkage between rational constructivist and spontaneous elements in the business firm's constitutional dynamics. Business firms as voluntary organisations embody much of the same dynamics as larger organisations such as nation states. On the other hand, it is clear that the interrelations among the members of business firms are distinguishable in many aspects from those among the members of a nation state. The broadly defined constitutional perspective can provide an explanation for institutional change within firms without introducing non-individualistically definable efficiency criteria, or without assuming away the subjective elements by referring to natural selection.

A central reason for my attempt to extend the constitutional approach to the firm by ways of conventions is based on the logic of reasoning. An analogy between a voluntary market exchange and a voluntary exchange of commitments in agreement upon a common rule is found problematic because the efficiency consideration of the latter involves the very rule that is yet to be agreed upon. A market exchange can be viewed as efficient within the rules that are already in place, whereas efficiency of an exchange that itself produces a shared rule cannot be derived from the exchange itself. Therefore, connecting a social contract with the conventions which define the shared understanding of the boundary between acceptable and unacceptable helps us to understand where the source of efficiency of an exchange upon shared rules is located.

The types of constitutional approaches to the firm, as represented in Gifford (1991) and Furubotn (1988) are seen to be slightly problematic. Gifford's treatment assumes epistemic capabilities on the part of the manager when designing an *efficient constitution* that may fall short in real-life contexts. The present approach is equally sceptical about Furubotn's idea that a firm's constitution is a device by which profit maximisation *per se* is secured. It is maintained here that a social contract can only ensure that the mutual interests of the participants are recognised at the *moment of choice*. Whatever *consequences* such a contract will produce as the future is disclosed are beyond the epistemic limitations of the participants.

The contractarian principles are essentially about the means, not about the ends, by which a collective endeavour is to be pursued if the individual is taken as the ultimate source of valuation. The normative content of constitutional philosophy does not carry from this normative individualist position to consequential assessment of utilities that are not present when a choice is made. This chapter has discussed some central complications that such a position necessarily brings with it. Shared understanding of appropriate modes of behaviour in the form of conventions is proposed to help us in understanding the interplay between evolution and design in the constitutional dynamics of business firms.

## REFERENCES

Alchian, A. A. and H. Demsetz (1972), 'Production, Information Costs and Economic Organization', *American Economic Review*, **62** (5): 777–795.

Baker, G., R. Gibbons and K. J. Murphy (1997), 'Implicit Contracts and the Theory of the Firm', *NBER Working Paper*, No. 6177, Cambridge, MA: NBER.

Barnard, Chester (1938), *The Functions of the Executive,* Cambridge, MA: Harvard University Press.

Barnard, Chester (1976), 'Foreword', in Herbert A. Simon, *Administrative Behavior – A Study of Decision-Making Processes in Administrative Organizations*, Third edition, New York, NY: The Free Press, pp. xliii–xlvi.

Barry, Norman P. (1984), 'Unanimity, Agreement, and Liberalism. A Critique of James Buchanan's Social Philosophy', *Political Theory*, **12** (4): 579–96.

Block, W. and T. J. DiLorenzo (2000), 'Is Voluntary Government Possible? A Critique of Constitutional Economics', *Journal of Institutional and Theoretical Economics*, **156** (4): 567–82.

Brennan, Geoffrey and James M. Buchanan (1985), *The Reason of Rules,* Cambridge: Cambridge University Press.

Buchanan, James M. (1969), *Cost and Choice. An Inquiry in Economic Theory*, Chicago, IL: Markham.

Buchanan, James M. (1977), *Freedom In Constitutional Contract. Perspectives of a Political Economist*, College Station, TX: Texas A & M University Press.

Buchanan, James M. (1991), *The Economics and the Ethics of Constitutional Order,* Ann Arbor, MI: University of Michigan Press.

Buchanan, James M. and G. Tullock (1962), *The Calculus of Consent – Logical Foundations of Constitutional Democracy,* Ann Arbor; MI: University of Michigan Press.

Coase, R. H. (1937), 'The Nature of the Firm', *Economica,* **4** (16): 386–405.

Coleman, James S. (1990), *Foundations of Social Theory*, Cambridge, MA: Belknap Press.

Constant, Edward W. II (1980), *The Origins of the Turbojet Revolution*, London: Johns Hopkins University Press.

Foss, Nicolai J. (ed.) (1999), *The Theory of the Firm: Critical Perspectives*, London: Routledge.

Furubotn, E. G. (1988), 'Codetermination and the Modern Theory of the Firm: A Property-Rights Analysis', *Journal of Business*, **61** (2): 165–81.

Gauthier, David. (1998), 'David Hume, Contractarian', in David Boucher and Paul Kelly (eds), *Social Justice*, London: Routledge, pp. 17–44.

Gifford, A. Jr. (1991), 'A Constitutional Interpretation of the Firm, *Public Choice*, **68** (1–3): 91–106.

Granovetter, M. (1985), 'Economic Action and Social Structure: The Problem of Embeddedness', *American Journal of Sociology*, **91** (3): 481–510.

Hayek, Friedrich A. von (1952), *The Sensory Order. An Inquiry Into the Foundations of Theoretical Psychology*, London: Routledge & KeganPaul.

Hayek, Friedrich A. von (1973), *Law, Legislation and Liberty, Vol. 1, Rules and Order*, Chicago, IL: University of Chicago Press.

Hirschman, Albert O. (1970), *Exit, Voice, and Loyalty: Responses to Decline in Firms, Organisations, and States*, Cambridge MA: Harvard University Press.
Hobbes, Thomas [1654] (1996), *Leviathan*, Oxford: Oxford University Press.
Kreps, David M. (1990), 'Corporate Culture and Economic Theory', in J. E. Alt and K. A. Shepsle (eds) *Perspectives on Positive Political Economy*, Cambridge: Cambridge University Press, pp. 90–143.
Lachmann, Ludwig M. (1976), 'On the Central Concept of Austrian Economics: Market Process', in Edwin G. Dolan (ed.), *The Foundations of Modern Austrian Economics*, Kansas City, MO: Sheed & Ward, pp. 126–32.
Langlois, R. N. (1995) Do Firms Plan? *Constitutional Political Economy*, **6** (3): 247- –61.
Milgrom, Paul and John Roberts (1992), *Economics, Organization and Management*, London: Prentice-Hall.
Nozick, Robert (1974), *Anarchy, State, and Utopia*, New York, NY: Basic Books.
Rawls, John (1971), *A Theory of Justice*, Cambridge MA: Harvard University Press.
Schlicht, Ekkehart (1998), *On Custom In the Economy*, Oxford: Clarendon Press.
Selznick, P. (1948), 'Foundations of the Theory of Organization'. *American Sociological Review*, **13** (1): 25–35.
Simon, H. A. (1951), 'A Formal Theory of the Employment Relationship', *Econometrica*, **19** (3): 293–305.
Simon, Herbert A. (1976), 'From Substantive to Procedural Rationality', in Spiro J. Latsis (ed.) *Method and Appraisal in Economics*, Cambridge: Cambridge University Press, pp. 129–48.
Sugden, R. (1993), 'Normative Judgments and Spontaneous Order: The Contractarian Element in Hayek's Thought', *Constitutional Political Economy*, **4** (3): 393–424.
Vanberg, Viktor (1985), 'Liberty, Efficiency and Agreement: The Normative Element in Libertarian and Contractarian Social Philosophy', Center for the Study of Market Processes, Working paper series 1985—18, George Mason University, USA.
Vanberg, Viktor (1986), 'Individual Choice and Institutional Constraints: The Normative Element in Classical and Contractarian Liberalism', *Analyse & Kritik*, **8**: 113–49.
Vanberg, Viktor (1992), 'Organizations as Constitutional Systems', *Constitutional Political Economy*, **3** (2): 223–53.
Vanberg, Viktor (1994), *Rules and Choice in Economics*, London: Routledge.
Vanberg, Viktor and J. M. Buchanan (1986), 'Organization Theory and Fiscal Economics: Society, State, and Public Debt', *Journal of Law, Economics, and Organization*, **2** (2): 215–27.
Wernerfelt, B. (1997), 'On the Nature and Scope of the Firm: An Adjustment Cost Theory', *Journal of Business*, **70** (4): 489–514.
Williamson, O. E. (1971), 'The Vertical Integration of Production: Market Failure Considerations', *American Economic Review*, **61** (2): 112–23.
Williamson, O. E. (1991), 'Comparative Economic Organization: The Analysis of Discrete Structural Alternatives', *Administrative Science Quarterly*, **36** (2): 269–96.
Wolff, Rirgitta (1997), 'Constitutional Contracting and Corporate Constitution', in Arnold Picot and Ekkehart Schlicht (eds), *Firms, Markets, and Contracts:*

*Contributions to Neoinstitutional Economics*, Heidelberg: Physica-Verlag, pp. 95–108.

# 8. Science as a spontaneous order: an essay in the economics of science

**William N. Butos and Roger Koppl**

---

> Perhaps it is only natural that in the exuberance generated by the successful advances of science the circumstances which limit our factual knowledge, and the consequent boundaries imposed upon the applicability of theoretical knowledge, have been rather disregarded. It is high time, however, that we take our ignorance more seriously. (Hayek 1967b)

## 1. INTRODUCTION[1]

What drives and influences scientific development? The last century saw increasingly subtle and complex answers to this question. We will propose a very broad model of science as a 'spontaneous order'. We believe our model helps to explain the move toward complexity in 20th century philosophy of science. It may also help resolve some of the difficulties into which students of philosophy and 'science studies' have been led. A brief and selective tour of the philosophy of science sets the context for our model.

Our question is: What drives and influences scientific development? The answer of traditional philosophy of science was often 'rationality'. Science was viewed as the apotheosis of rationality. In the 20th century, logical empiricism was one of the highest expressions of this tendency. The logical empiricists were positivists and heirs to the slightly earlier tradition of logical positivism. A leader of this group, Otto Neurath, celebrated 'modern attempts to reform generalization, classification, testing, and other scientific activities, and to develop them by means of modern logic' (Neurath [1938] 1955, p. 10). The very term 'logical empiricism', Neurath explains, 'expresses very well', the 'synthesis' between the 'empirical work of scientists' and 'the logical constructions of *a priori* rationalism'. He proudly claims, the two have now become synthesized for the first time in history' (Neurath 1938, p. 1).

Karl Popper is famous for his attack on positivism. The criticisms he and others offered led to the demise of positivism, at least in philosophy of science. Bruce Caldwell draws our attention to Frederick Suppe's remark, made in 1977, that 'positivism today truly belongs to the history of philosophy of science' (Suppe 1977, p. 632 as quoted in Caldwell 1982, p. 37).

Popper's own 'falsificationism' did not hold the centre. In 1962, Thomas Kuhn published his famous essay, *The Structure of Scientific Revolutions*. Science is not governed by rational procedures ensuring the gradual accumulation of true knowledge. Science is a social process that proceeds in fits and starts. Calm periods of 'normal science' are interrupted by the 'revolutions' of 'extraordinary science'. Kuhn's portrait of science seemed to challenge forcefully the old view that science is rational. It was the main progenitor of the 'growth-of-knowledge tradition' of Lakatos, Feyerabend, and, others.[2]

The growth-of-knowledge tradition is heterogeneous. Nevertheless, some common threads exist. Scholars in the growth-of-knowledge tradition emphasise the evolution of science as a cultural phenomenon. They look more closely at scientific processes than at any equilibrium theories or methods to which such processes might lead. They give greater emphasis to social and institutional issues. According to Caldwell (1982, p. 90), 'Their most wide-ranging break with the earlier tradition lies in the suggestion that the scope of philosophy of science be greatly enlarged.' With this break, 'Critical roles for the history, sociology, and even psychology of science emerge.'

In his 'Dilettante's Review' of the growth of knowledge literature, Caldwell (1982, pp. 89–93) noted the open-ended nature of this tradition. 'Each philosopher has a particular image of how science evolves' and what constitutes 'progress'. These competing views 'raise questions over the proper definition of such key terms as rationality and progress, questions which have yet to be successfully resolved' (Caldwell 1982, p. 91). Indeed, with the addition of voices such as that of Bloor (1976) and Latour and Woolgar (1979), Caldwell's questions have grown even more open and contentious. The history, sociology, psychology, and now economics of science have taken an increasingly important role in the study of science and scientific method.

Today, the rational elements emphasised by the logical empiricists and the extra-rational, evolutionary elements emphasised by Kuhn are mixed in different combinations by different thinkers. No consensus view seems to have emerged; no conventional wisdom exists. Today's competing visions are characterised, however, by a growing appreciation of the great complexity of science and scientific standards. This growing recognition is the culmination of a long-term trend.

The logical positivists had a relatively simple vision of science as verifiable knowledge. The logical empiricists recognised the oversimplifications of the logical positivists and moved to a more complicated account. The new philosophy recognised that not every scientific statement is subject to independent test and that some scientific terms cannot be expressed in an 'observation language'. Popper crafted a more complex account of science, in which falsifiability was the demarcation criterion. In Popper's account, science is a fallible process of error elimination in which the truth is approximated increasingly well over time. Kuhn's famous essay leapt to a markedly higher level of complexity. In addition to 'revolutions', 'normal science', and 'extraordinary science', we have scientific communities whose social structure influences the evolution of knowledge. Writers since Kuhn have woven even more intricate and nuanced tales of scientific process. Some of them lean more toward old-fashioned ideas of objectivity and rational procedure. Others lean more toward new-fangled notions such as truth being socially constructed or scientists being socially determined. All are rich with institutional and historical details.

The competing tendencies of the current situation have resulted in two extreme views, which represent the poles of a continuum along which most views fit reasonably well. We might call the one extreme 'traditional' or 'modern', and the other 'postmodern'.

In traditional interpretations, science is like a police report. The witness is interrogated and the truth is revealed. The most prominent advocate of this view is Francis Bacon, who extolled the 'prudent interrogation' of nature (Bury 1955, p. 57). Like a hard-nosed detective getting to the bottom of things, Bacon thought he was building 'a true model of the world, such as it is in fact' (Bacon 1955, p. 534). Like careful and precise police work, Bacon's task 'cannot be done without a very diligent dissection and anatomy of the world'. Men must 'begin to familiarise themselves with the facts' (Bacon 1955, p. 468). Bacon wants just the facts, maam. The substance of Bacon's position never really caught on. But its spirit dominated modernist philosophies of science such as logical positivism and logical empiricism.

In more recent interpretations, science is like a detective novel. The author invents the crime that is solved. Bruno Latour (1987) may be the most prominent advocate of this view. Latour and Woolgar (1979) insist that the 'out-there-ness' of 'reality' is 'a consequence of scientific work rather than its cause' (Latour and Woolgar 1979, pp. 180–2 as quoted in Kitcher 1993, p. 165). It makes perfect sense to ask if the butler did it. The reality created by the novelist exculpates some, implicates others. We cannot have just anyone found out. But the author constructs the governing reality that implicates the butler. We cannot appeal to 'external reality' to criticise the novel because the novel creates its own reality. The view that reality and truth are constructed is often called 'postmodern'.

The modern and the postmodern are in opposition. Is science a police report or a detective novel? For the modernist, science is based on fact. For the postmodernist, it's social causation all the way down. All sorts of compromise views have been proposed, stretching out between the modern and postmodern poles. The extreme modernist pole has been pretty well abandoned. Everybody agrees that facts are theory-laden. We know that social influences operate on scientists and that the data under-determine theory. We all agree that 'rationality' is a difficult notion. The postmodernist pole is well populated. Those who eschew that pole are engaged in an essentially conservative and rearguard action. Kitcher, for instance, is explicit in his attempt to strike a compromise between older and newer views. The older view, Kitcher (1993, p. 390) says, 'does not require burial, but metamorphosis'. We prefer to get off the road between these poles and try out the rather different view of science as a 'spontaneous order'. And, as we indicate below, in applying this view to science we shall also provide some perspective on how scientific knowledge is generated.

In society, a 'spontaneous order' is a structure that evolves and persists as an unintended consequence of human action. Science is not composed by an author, as is a novel or police report. Like the common law, it emerged as an unintended consequence of human action. As with litigants and lawyers, scientists are driven by many motives, many of which are low. In Anglo-American jurisprudence the interaction of many persons has created the principles of common law, an emergent order that guides the actions of those who come later. Science is governed by a similar set of evolved principles. In science and jurisprudence, participants are animated by diverse goals and errors are common. We can nevertheless ask if the system, science or law, tends to produce good decisions and if it beats the available alternatives.

Our view draws on the work of Hayek who explained minds, markets, and societies as spontaneous orders (see for example Butos and McQuade 2001, Gray 1984, Lavoie 1985). We believe science can be analysed as an emergent or 'spontaneous' order. But that is not to deny that science is partly spontaneous and partly designed. But the design elements, we suspect, have been exaggerated in the past. With some exceptions (such as Polanyi 1962, Wible 1998), the spontaneous elements have been underplayed. If we are right about that, the one-sided view expressed in this chapter may add something to the conversation.

Our view of science as a spontaneous order leads us to a middle position on many of the issues separating the modern and postmodern views. But ours is not a compromise between two extreme views on a continuum. It is based on a different set of analytical principles derived from our treatment of science as a spontaneous order. Our perspective makes scientific rationality a product of social process. We do not have an under-socialised view of rationality. We would not claim, therefore, that 'neoclassical' rational choice models are a complete and adequate picture of scientific activity. We do not however discard the notion that standard economics is a potentially useful analytical engine. We do, however, go beyond rational choice models by recognising science as 'processive' and rationality as, in part, tacit.

The concept of spontaneous order is similar to the concept of 'complex adaptive system' from the complexity theory of the Santa Fe Institute. Stuart Kauffman (1993) uses the term 'spontaneous order' without citing Hayek or, apparently, knowing of his existence. Complexity theorists have constructed mathematical models of science as a complex adaptive system. As far as we can tell, the Santa Fe vision is complementary to that of Hayek (Koppl 2000a; 2000b). In this essay, however, we raise points better represented in Hayek's work than in that of most complexity theorists. Our points are more philosophical than technical and are thus better expressed in ordinary language than in the special language of mathematics.

## 2. SCIENCE AS A SPONTANEOUS ORDER

A spontaneous order is a system whose order is not imposed from outside, but comes from within the system (Hayek 1973, p. 36). Not all spontaneous orders are social. There are spontaneous orders in nature, too. Biology may be the best source of examples. Hayek notes, 'biology has from its beginning been concerned with that special kind of spontaneous order which we call an organism' (ibid., p. 37). More recently, 'under the name of cybernetics', the physical sciences have produced 'a special discipline which is also concerned with what are called self-organizing or self-generating systems' (ibid., p. 37). Spontaneous social orders are a subset of spontaneous orders. Hayek's phrase, 'spontaneous order', has two elements; spontaneity, and order. Each term merits separate discussion before turning to the characteristics of spontaneous orders.

### 2.1 What 'Spontaneous' Means

What is 'spontaneous' in a spontaneous order is not the behaviour of the elements, but the overall order. In spontaneous orders, the order comes from within. It is a result of each of several elements behaving according to some set of rules. The regularity of the elements produces regularities of the system. The elements must be regular in their behaviour, whether that behaviour is 'spontaneous' or not. In the case of spontaneous social orders, it may be that the individuals whose actions generate the order are not 'spontaneous' at all. Perhaps their actions are all 'rational' in some sense. But the order will be spontaneous if the actions were not specifically designed to produce the overall order.

## 2.2 What 'Order' Means

Hayek's definition of order is very broad. We think it may be too broad to be of use in the study of spontaneous orders. For Hayek, a group of events falls into an 'order' when knowledge of a part permits one to infer something about the rest. Events form an order when they form a pattern or set of regularities.[3] This definition seems to cover all cases of order. It suggests, however, a surprisingly passive view of what spontaneous orders do and how they function.

Hayek's own writings point to a more interesting and fertile way to understand the concept of 'order', at least in the context of self-organisation. The sort of orders he describes in his economics, his political theory, and his philosophy of mind are adaptive orders. The notion of 'adaptation' is fundamental to the complexity theory of Holland (1992), Kauffman (1993), and other theorists of 'Santa Fe complexity'. In the complexity literature, an order or system may be 'adaptive' in either of two senses. First, an order is adaptive if it changes to better fit its environment (Holland 1992, p. 28). Second, an order is adaptive if it contains elements that are adaptive in the first sense.[4]

Adaptive orders in biology, Kauffman argues, 'must "know" their worlds'. They 'sense, classify, and act upon their worlds'. Organisms 'have internal models of their worlds which compress information and allow action'. His examples include '*E. coli* swimming upstream in a glucose gradient', and 'a tree manufacturing a toxin against a herbivore insect' (Kauffman 1993, p. 232). In the case of *E. coli*, Kauffman says, 'I permit myself the word "classified" because we may imagine that the bacterium responds more or less identically to any ligand binding the receptor, be it glucose or some other molecule' (ibid., p. 233). Like Kauffman and other complexity theorists, we recognise a connection between adaptation on the one hand, and the generation of knowledge through classification and reclassification on the other hand.

By 'adaptive order', we mean a structure, constituted by interacting elements connected to each other via routines, which is capable of producing a 'classification' over some defined domain or environment (Butos and McQuade 2001; McQuade and Butos 2002). The classification that an order produces depends on the characteristics of its constitutive elements and the various ways they interact. This classification is likely itself to be multi-dimensional, requiring any number of different ways of scientifically understanding the order. We might, for example, describe an order by its morphology, perhaps at different points in time. If our interest were oriented toward the order's functioning, its physiology and emergent characteristics would require special attention, including the order's capacity to generate classifications. But these emergent properties, representing the outcomes an order produces, constitute a kind of knowledge, that is, a classification, by which we come to understand the order. Because orders differ in their complexity, in terms of their constitutive elements (and their interactions) as well as their capacities or opportunities to interact with each other, the knowledge- or classification-generating capacities of orders can be expected to differ in relevant and interesting ways.

Hayek (1952a), for example, comprehends 'mind' as a particular order in which the connections between neurons result in the transformation of 'raw' sensory impulses into knowledge (a classification) about aspects of the individual's environment and external reality. According to Hayek, the operation of the central nervous system involves the *generation* of knowledge of external reality, not simply a reproduction of disembodied external knowledge in a 'language' that as humans we can understand. Nor, for Hayek, is the mind to be viewed as a device for simply transmitting knowledge of external reality to us. Rather, according to Hayek, the mind is an active structure with the capacity to produce various emergent characteristics – and perhaps of greatest interest among those characteristics is the ability of the mind to generate knowledge and with it an interpretation of reality that results from the mind's structure and functioning.

It is tempting to suggest that in Hayek's cognitive theory a definite mapping could be discovered linking together the sensory inputs with a resulting and specific classification in the sense of finding an algorithm that would describe how certain inputs deterministically produced certain outputs. But there are both theoretical and practical reasons which impose certain restrictions on that sort of claim. First, the mind is a highly complex structure that is practicably very difficult to penetrate. Second, and as we will discuss below, Hayek proposes a 'limits to explanation' argument that establishes a non-specifiable but nonetheless binding theoretical constraint on any complex order to fully explain itself. But these arguments need not detain nor especially deter us from also emphasising that these practicable and theoretical limits to explanation cannot tell at what specific point they become operative.

Of special interest to economists is the particular social order of the market – here understood as a structure composed of individuals who interact (exchange goods and services) according to certain rules concerning the use and disposal of property rights (Hayek 1973). The market order, or catallaxy, bears important functional similarities to the mind. Like the mind, the interactions constituting market exchanges transform the knowledge (consciously and tacitly) held by individuals into a different kind of classification or knowledge: market prices and quantities and goods-characteristics. The knowledge so generated does not, of course, emerge apart from the interactive behaviours of individuals and the arrangements and mechanisms by which the order operates; consequently, the kind of knowledge the catallaxy produces is not a straightforward aggregation of the (human) knowledge each individual has or that of a 'collective consciousness'. Indeed, it is precisely the recognition that the catallaxy functions without a central locus of control that looms so large in our understanding of the order. What we can say is that a catallaxy produces a particular kind of classification – a market price – as a unique outcome of its operation and that such knowledge is generated as a by-product of the interactions constituting the operation of the order. This suggests that we cannot anthropomorphise the market or commit what Hayek identifies as the 'rational constructivist' fallacy.

However, various non-catallactic orders also exist as identifiable phenomena within the social realm. These orders, though functioning in the context of the market economy and often having close ties to it, nonetheless represent structures that differ in important ways from those of the catallaxy, particularly in terms of the mechanisms that regulate and coordinate interactions and in terms of their classification or knowledge-generating capacities (Butos and Boettke 2002). Of interest in this chapter is the particular order identified as the 'scientific order'. The elements of this order are scientists who interact in a variety of ways (such as publications, conferences). These interactions generate particular classifications that we can simply refer to as 'reliable and codifiable knowledge' about reality. In science, we often find that accepted conventions governing interactions among scientists take the form of 'procedures' thought to be useful in generating knowledge (or weeding out the relatively blacker claims). But acceptable procedures are also an emergent property of the order itself and so may be expected to change. The motivations of individual scientists are not crucial for the outcomes generated by science: we can suppose some scientists motivated by lofty ideals or by baser ones. What matters is that a specifiable context enables scientists to interact in particular ways and that in so doing, an order is formed that generates as a by-product of those interactions a classification we designate as 'scientific' or 'reliable and codifiable' knowledge.

In suggesting that science is a 'spontaneous order', we wish to note that its emergent characteristics arise because its elements are following rules of action and interaction and not because these members are being placed into patterns under the direction of a guiding mind. A catallaxy is a spontaneous order; a Japanese rock garden is not. 'Invisible-hand explanations' are meant to explain spontaneous orders of human action. In society, spontaneous orders are 'the result of human action, but not of human design' (Hayek 1967c). (See Koppl 1992 for a discussion of invisible-hand explanations.)

### 2.3 The Fundamental Distinction Between Spontaneous Order and Organisation

Hayek reports that after completing *The Constitution of Liberty*, he was led to write the three volumes of, *Law, Legislation and Liberty* by the 'recognition that the preservation of a society of free men depends on three fundamental insights which have never been adequately expounded and to which the three main parts of this book are devoted' (Hayek 1973, p. 2). The first volume, *Rules and Order* was devoted to the first of these fundamental insights, namely, 'that a self-generating or spontaneous order and an organisation are distinct, and that their distinctiveness is related to the two different kinds of rules or laws which prevail in them'.[5]

In Hayek's account, there are three 'distinguishing properties' of spontaneous orders not shared with organisations. These distinguishing properties must be emphasised because 'we tend to ascribe to all order certain properties which deliberate arrangements regularly, and with respect to some of these properties necessarily, possess' (1973, p. 38). The three distinguishing properties of spontaneous orders are the opposites of the properties we tend to ascribe to organisations. First, organisations 'are relatively *simple* or at least necessarily confined to such moderate degrees of complexity as the maker can still survey'. Second, 'they are usually *concrete* in the sense just mentioned, that their existence can be intuitively perceived by inspection'. Finally, 'having been made deliberately, they invariably do (or at one time did) *serve a purpose* of the maker' (ibid., p. 38).

The 'degree of complexity' in a spontaneous order 'is not limited to what a human mind can master. Its existence need not manifest itself to our senses but may be based on purely *abstract* relations which we can only mentally reconstruct. And not having been made it *cannot* legitimately be said to *have a particular purpose*, although our awareness of its existence may be extremely important for our successful pursuit of a great variety of individual purposes' (1973, p. 38). Thus, spontaneous orders tend to be purpose-free, abstract, and complex. Any discussion of science as a spontaneous order must begin with a discussion of these three characteristic features of spontaneous order. Science, we argue, exhibits all three characteristic features of spontaneous orders.

### 2.4 Science Has No Overall Purpose

An organisation has a purpose, or at least began with one. A spontaneous order has no purpose. If science is a spontaneous order, then it, too, has no overall purpose. The existence of science may promote many individual purposes. The existence of science may serve a useful function in society. But science itself is an ends-independent enterprise.

What unites the various activities of scientists under the umbrella of science is not a unitary hierarchy of values, but a shared set of rules. In idealised versions, these rules require that one report only on replicable experiments, that one not falsify data or hide seemingly contradictory evidence, that one wash test tubes thoroughly and keep measuring instruments in good repair, and so on. But that is like saying the rules of common law require one to keep one's promises, bargain fairly, and never, never cheat. In reality, the official rules of science are loosely related to the rules that really guide scientific decision-making just as the official rules of common law are only loosely related to the rules that really guide contracting parties. In both cases, the real regularities are more complex and less charming to behold than the official rules. But if science is indeed a spontaneous order, the actions of scientists are rule-governed activities aiming at various particular ends. Just as the rules of common law exist and prevent us from many undesirable acts, the rules of scientific behaviour exist and constrain behaviour. There is no grand overseer directing all the efforts of the world's scientists. However much science may be like or unlike the market for leather shoes, scientists are engaged in rivalrous competition with one another. And if they are engaged in rivalrous competition, their ends are competing and, to some extent at least, incompatible. Scientists are united by means and not ends.

The rules governing a spontaneous order differ from the rules governing an organisation. The purposeful character of an organisation requires that the rules constraining its members be 'rules for the performance of assigned tasks'. These rules 'are thus necessarily subsidiary to commands, filling in the gaps left by the commands' (Hayek 1973, p. 49). The rules of an organisation are ends-dependent. Those of a spontaneous order are ends-independent.

The ends-independent character of the rules of science explains their resemblance to ethical rules. The rules of ethics are rules for getting along in a spontaneous order. Ethical rules have a real role to play in guiding individual behaviour and our judgments of the behaviour of others.

Nevertheless, the system is complex. Ethical norms are a kind of publicly recognised ideal violated in varying degrees in daily practice. It is probable that a strict universal adherence to our stated ethics would reduce social cooperation.

The rules governing scientific activity are similar to the ethical norms of social life. Thus, it was reasonable for Merton to speak of 'the ethos of science'. In Merton's analysis, scientists internalise in varying degrees four fundamental scientific values. The value of universalism holds that we do not reject theories because of the race or nationality of their authors (Merton 1957, pp. 553–6). The value of 'communism' holds that the results of scientific study are the common property of all (ibid., pp. 556–8). 'Disinterestedness' holds that we set aside our own interests and desires in judging the results of our own and of others' work (ibid., pp. 558–60). Finally, science is, for Merton, 'organised scepticism'. We must withhold our judgment until the facts come in and we must maintain always the attitude of 'detached scrutiny' of the evidence.

Like many contemporary perspectives, the view that science is a spontaneous order helps us to maintain a sophisticated understanding of Merton's four norms. Merton himself is less naïve than he has sometimes been portrayed. Of disinterestedness, for instance, he says that it is not a matter of the superior morality of scientists. 'It is rather a distinctive pattern of institutional control of a wide range of motives which characterises the behaviour of scientists'. Scientists are not unusually ethical; rather, 'the activities of scientists are subject to rigorous policing' (1957, p. 559). Nevertheless, he seems content to present the official or idealised rules of science and ask what institutional and social structures promote the scientific ethos. Merton's norm of 'organised scepticism', for example, neglects the role of faction and commitment to which Kuhn (1970, pp. 40–2, 176–81) and others have drawn our attention. (Merton also seems to conceive of science as an organisation; cf. Merton 1957, p. 552.) We prefer a relatively sceptical view of the Mertonian norms. We prefer to recognise and explore the complex dynamic between the (often tacit) rules by which individual scientists are guided in their daily practice and the evolving public norms to which they pledge allegiance.

### 2.5 Science is Abstract

If science is a spontaneous order, it is composed of meaningful actions and their (often unintended) consequences. The elementary 'facts' of the phenomena under study are more or less familiar to us. What need scientific elucidation are the connections between such actions and their unintended consequences. In the study of science, as in all of the social sciences, there is a kind of reversal of the situation characteristic of the natural sciences. In the natural sciences we observe an overall order and make conjectures about the underlying phenomena that fit together to produce the order. In the social sciences, by contrast, we are more familiar with the elements that together generate an order, even though the order itself cannot be directly observed. We do not observe, say, the market order in the same way we observe the paths of the planets. We imaginatively reconstruct the order out of our knowledge of the characteristics of the actions that produce it (Hayek 1952b, pp. 65–7). In the economics of science, too, we are more familiar with the component scientific actions of talking, calculating, and so on, than with the overall order generated by them.

This distinction between the natural and social sciences is a matter of degree. In both fields there is a kind of movement between the elements and the overall order they produce. In natural science our pre-scientific account of the observed order is replaced, through scientific explanation, by another, more reliable description. The order we explain is not quite the same as that which we set out to explain. In the social sciences we start out with some knowledge of the overall order produced by the many actions by which we explain it. Paris is fed. But it seems fair to say that in social science the elements are more familiar than the order and in natural science the order is more familiar than the elements.

## 2.6 Science is Complex

If science is a spontaneous order, then it is complex. Hayek measures the degree of complexity of 'different kinds of abstract patterns' by the 'minimum number of elements of which an instance of the pattern must consist in order to exhibit all the characteristic attributes of the class of patterns in question' (Hayek 1967b, p. 25). This is similar to John Maynard Keynes's notion of the 'degree of independent variety' (Keynes [1921] 1973, p. 279). Complex phenomena have many determinative 'elements' or 'degrees of independent variety'. As we noted earlier, the idea is also similar to the 'complex adaptive systems' of the Santa Fe Institute.

If science is a complex phenomenon, the study of science must confront some essential, logically implied, limits to explanation.[6] A complete explanation of science would be a meta-theory. The meta-theory would include a representation of the object theory and thus of the phenomena explained by the object theory. It would also need an independent model of the phenomena explained by the object theory. The meta-theory would require, further, a model of how the explananda of the object theory – as represented in the meta-theory – give rise to the object theory. This model would necessarily include a model of the scientists who produce and reproduce the object theory. The job of the producer of the meta-theory is thus more complex than that of the scientists whose ideas she or he would like to give a causal explanation of. If the meta-theory were scientific, it would apply to itself, in which case it would be more complex than itself! If Hayek was right to deny that we have the wits to perform a radical critique of society, then we lack equally the wits to perform a radical critique of science. We lack, in short, not only an Archimedean lever to design society, but also one to design science.

If science is a complex spontaneous order, then the economics of science can say only so much about it. In that case, it would seem to be one of the tasks of the economics of science to seek out and define its own limits of application. While the discovery of limits may seem a gloomy affair, it has advantages. If we choose to view science scientifically, there arises the problem of reflexivity (Bloor 1976, pp. 13–4; Pickering 1992, pp. 18–22). (We will return to this issue in the next section.) The argument that complexity implies limits of explanation reduces the potential for reflexivity to create paradox. A scientific explanation of a complex phenomenon must restrict itself to an explanation of the principle (Hayek 1952a, pp. 182–4; Hayek 1967a). We can only determine, therefore, what kinds of circumstances produce what kinds of science. Specific causes for specific beliefs will not, in general, be capable of discovery. Thus are we spared the embarrassment of a theory that, when turned on itself, explains itself away.

A simple model of explanation may illuminate the issues. (We call this the 'counting argument'.) The model is an adaptation of arguments found in Hayek's work (1952a, pp. 184–90; 1967d, pp. 60–3). The arguments are central to Hayek's theory of spontaneous orders. As we have seen, in Hayek's theory, organisations are relatively simple. The order can be surveyed by an individual mind. This simplicity is the necessary consequence of design. Somebody arranged an arranged order. To do his or her work, the author must survey the scheme. Thus, organisations are relatively simple. Precisely because they were not designed, spontaneous orders may be more complex than an individual mind can survey. The counting argument, and the arguments of Hayek from which it is derived, show that the greater complexity of spontaneous orders creates logically necessary limits to our understanding of them. In the present context, that means logically necessary limits to our understanding of science.

Any theory will entail a description of some sort of context, environment, or situation. Within this situation (as we will call it from now on) will be certain objects whose behaviour depends on the situation. The objects may be persons, atoms, or ecosystems. Changes in the situation may be changes in one or more exogenous variables. Changes in the objects may be changes in one or more endogenous variables.

The theory will define the situation by indicating the values of several variables or dimensions. There will be, say, N dimensions that go into the theory's description of a situation. For the Marshallian cross, these are the prices of related goods, income of demanders, and so on as well as technology, factor prices, and so on. Each of the dimensions of the situation can take on different values. Income can rise, fall, or stay the same; factor prices may rise, fall, or stay the same; and so on. To keep our notation simple, we will assume each dimension can take on the same number, M, of different values. Thus the theory will be able to distinguish $M^N$ different situations. The number of objects or object types is S. In the Marshallian model of supply and demand, there are two object types, suppliers and demanders. Each object type may behave in several different ways. Buyers may attempt to purchase one unit, two, or more. Again, for the sake of simple notation we assume this number, R, is the same for all objects. Thus the constellation of objects may take on $R^S$ different configurations.

A theory with these characteristics can generate at least $(R^S)(M^N)$ separate statements. Each of these statements describes one state of the environment, one configuration of the constellation of object types within the environment, and an indication showing if, for that pair, the two correspond or not.

Note that $(R^S)(M^N) > R^S > R$. This implies that the theory must be more complex than the phenomena it explains, and more complex than any of the object types it identifies. This greater complexity refers not to the objects themselves, but to the theory's representation of the object types. The whole theory is more complex than a part of it.

This simple calculus, it may be objected, depends on the assumption that R, S, M, and N are finite numbers. But there are typically an infinite number of behaviours an object in our model might take on. This objection is mistaken. Our theory may contain statements such as 'Agents choose a probability, p, between zero and one half'. This statement seems to imply that any agent's behaviour may take on a continuum of values. But our theory may only distinguish a few cases. We may want to know only if p is 'sufficiently large' for some condition to hold. Consider the Marshallian model of supply and demand. Each demander is seemingly free to buy any of a continuum of units. But our description really only distinguishes quantities greater than, less than, and equal to those of the initial equilibrium. Far from a continuum of actions, we have only three. This is typical. Any model distinguishes only a finite number of possibilities. This point may be illuminated by a thought experiment.

Imagine we have a complete formal language for the theory in question. (Carnap (1958) provides examples for physics and biology.) Such a language would be composed of a finite number of symbols and a finite number of rules for combining the symbols. The rules determine what are well-formed formulae. The number of well-formed formulae is necessarily countable. If we imagine *any* finite limit to the length of a formula, then the language can distinguish only a finite number of cases. Given any amount of time required to write a symbol, however brief, one can calculate a finite formula length so great that it would require more time to write the formula than has yet passed in the history of the universe. Thus whether or not one can calculate a least upper bound to formula length, an upper bound exists. It seems reasonable to conclude that only finite numbers of states and behaviours can be distinguished by any theory.

In the social sciences, one or more of the S objects may be an 'ideal type', the representative agent in a New Classical model for instance. In that case, the ideal type, the scientist's model of S, is necessarily simpler than the larger model within which S acts. If the ideal type is simpler than the scientist's model, he or she must be simpler than the scientist. But if the real persons who correspond in some way to the ideal types are not simpler than the scientist, then the ideal type is a very incomplete picture of them. It refers only to some general features of the agent's actions. Whether or not the objects are ideal types, if the system under study is complex, the model will explain only general features of it. This is what Hayek means by 'explanation of the principle'. Our representations of explained phenomena must always be simpler than the model or theory with which we explain them. The representation is just a part of the larger model that explains the phenomena. Our representation of any object must, therefore, be simpler than we are.

When we engage in explanations of the principle, we remain silent about many particulars. When we try to test a relationship of which we have knowledge of the principle only, there is a gap between the relative specificity of the test case and the general relationship under scrutiny. The gap is filled by 'maintained hypotheses'. Any apparent falsification may be blamed on the failure of a maintained hypothesis, rather than the falsity of the theory.

There are limits to explanation in the study of science. Causal accounts can be given, but they will generally be explanations of the principle. The causal links we can reasonably hope to establish will probably be found in models with only very general ('thin') descriptions of scientific activity. The descriptions amply endowed with many particulars ('thick' descriptions) will be less able to sustain a satisfying causal account. Causal stories will not give a complete account of particular episodes in the history of science. It is a good rule of thumb that thick description is suited to history while theory relies on thin description. (This point is carefully developed by Machlup (1936) who cites Schutz ([1932] 1967.)

If we are right about the limits to explanation in the study of science, then we can understand why methodology is somehow both ever-important and yet ever-over-reaching. A full-blown prescriptive methodology is indeed out. But methodological principles, rooted in practice, may often be articulated. This is analogous to the jurist's work.

If science is a spontaneous order, the order of which is imaginatively reconstructed from its 'simplest elements', then empirical testing is very important to the study of scientific enterprise. It is not enough to report 'field observations'. One must have an engine that produces testable propositions about the nature of scientific order. Then one has grounds to say, depending on test results, this theory of science is better than that one.

## 3. IMPLICATIONS FOR THE RATIONALITY AND EPISTEMICS OF SCIENCE

The view of science as a spontaneous order contrasts with others. Notably, it is quite different from the whipping-boy of many writers, logical positivism. We want to apply our lashes as well, but not out of an excess of cruelty. We think the contrast will help to highlight some of the strengths of our position. In particular, the positivist view we take as whipping-boy and logical foil is a relatively clear example of the view that science must be algorithmic to be rational. Such an algorithmic view makes rather dull puppets of individual scientists. It is our impression that this view, in altered or incoherent form, still has some currency. Even though positivism is long dead, even among economists, it left behind some 'modernist' dispositions that we hope to combat in this chapter.

The view of science as inherited from logical positivism and (some versions of) logical empiricism centred on the meaning and truth status of propositions. Their aim was to provide criteria to distinguish scientific from non-scientific propositions. This is the 'demarcation problem'. The underlying epistemology of positivism was justificationism. Propositions were deemed true if they were justified. Justified propositions conveyed true and indisputable knowledge. They had been verified or confirmed by adherence to some set of procedures. These procedures were based on the supposedly ultimate epistemological authorities of sensation and rational thought. A standard problem for any justificationist system is the difficulty in justifying its final epistemological authority. To avoid an infinite regress one must 'retreat to commitment' (Bartley 1984) and cleave unquestioningly to some epistemological authority. Positivists retreated to sensation and rational thought. It is also clear that this sort of philosophy of science provided little real guidance for researchers. Rationality in science was equated with propositions that conveyed justified, proven knowledge; what it meant to be a rationally acting scientist did not enter the picture. Scientists were enjoined to be rational, but given little concrete guidance on what that meant in their daily practice.

Science progresses, according to this justificationist view, in a curiously mechanical and anonymous way. Inadequate attention was paid to an obvious fact, namely, that propositions are maintained as justified true belief by the actions of real-life, flesh-and-blood scientists. Scientists were wrongly viewed as props for the entirely logical task of empirical verification. The entire process is analogous to a kind of input-output machine or to a bureaucratic organisation in which means and ends are largely given and known. This view of things seems compatible with the creation of a centralised scientific bureau. If such a bureau could command the necessary inputs to algorithmically crank out true (justified) statements about reality, it would turn science from a spontaneous order into a centrally planned organisation, from an undesigned 'cosmos', as Hayek would put it, into a purpose-seeking 'taxis'. In unsophisticated hands, this positivist view makes science into some sort of grand algorithm and scientists into automata responsible for subroutines of the grand algorithm.

If science is a spontaneous order, the positivist picture of science (and others of a similar nature) is not sustainable. The view of scientific rationality consistent with the spontaneous-order view contrasts sharply with positivism. The epistemic implications of our view are very different too. We discuss each point in turn.

Science has no overall purpose. Science is ends-independent; it is open-ended and evolving. It does not move toward any fixed point or specific goal. It is 'processive', that is, characterised chiefly by the unfolding of a process that has no particular tendency to wind down and is not headed for any particular end point. Because science is processive and open-ended, it does not ordinarily produce final and definitive outcomes or standards. Everything is up for grabs; any theory may be overturned. This goes for both scientific theories and the standards of judgment scientists employ when choosing among theories. We do not mean to say that anything goes, no standards apply, and no results are reliable. We do mean to say, however, that no past achievement is immune from challenge and that surprising changes in both standards and results are not only possible, but likely. Such changes are characteristic of the process. Working scientists will generally be in a position to make competent local judgments about theories and standards. Every concrete judgment, however, will be formed by the contingencies of time and place. Global and timeless judgments constitute over-reaching.

The scientific order has no overall purpose. Individual scientists and institutions, however, do have purposes. If scientists and scientific organisations act reasonably in the pursuit of their goals, if they react appropriately to their situations, then the analysis of science would seem to be subject to the requirement of 'methodological individualism'. This, of course, does not mean that our understanding is strictly defined in terms of those separate individual purposes: methodological individualism does not imply epistemological individualism. Indeed, our aim is precisely to use a sensible methodological principle to say something about science as a social structure. The sort of methodological individualism we have in mind has been described as 'institutionalist individualism' and 'individualist institutionalism' (Langlois 1989). This brand of methodological individualism recognises the impact of the total situation on the individual and it recognises the role of the total situation in combining individual actions into aggregate outcomes. What makes it 'individualist' is its insistence that aggregate outcomes be traced to the individual actions that unintendedly generate the overall order (see Langlois 1989; Koppl 1994).

The order of science is abstract. The order we observe in a spontaneous order is not born of a central locus of control but of the interaction between the agents whose actions produce the order. A spontaneous order is not the product of any one single mind. Yet, spontaneous orders are determinate (not deterministic or predictable) in that they are governed by abstract internal regulative principles. The order we observe in spontaneous orders reflects the constraining effects of abstract or general rules. In the social domain, these rules provide the essential context of constraint within which the spontaneous order functions. For the actors, these rules are in many cases largely tacit and take the form of implicit prohibitions concerning what not to do. They are transmitted to students via customary ethics or training from mentors in the certification (educational) process. To a significant extent, rationality for the individual researcher means following certain general (abstract) principles of action. At the same time, however, it is precisely by following certain general rules and by not submitting in most cases to specific commands that the growth of knowledge occurs. Novelty is produced as an unplanned by-product of researchers following various general rules of sciences. These general rules create freedom within constraints. The scientific community, we might say, operates under a kind of ordered liberty.

Science is a complex phenomenon. The counting argument presented earlier suggests that inherent constraints exist on what can be known about complex phenomena. From that argument, which follows Hayek, we know that a model of a given degree of complexity can explain only phenomena of lesser complexity. Thus, a social structure (such as a scientific community) possesses a degree of complexity that exceeds that of any member (or observer) of that community.[7] Certain aspects or characteristics of that community are in principle closed to each member. Knowledge inheres within the community that is not available to any one person; in short, a division of knowledge exists among the members of that community. This is nothing other than the cross-sectional aspect of the more familiar idea that the accumulated knowledge of a scientific community exceeds that of any constituent member. The central implications of this are that science necessarily requires social structures for its continuation and that, like all social phenomena, its functioning depends on mechanisms and institutions that foster the production and distribution of knowledge not available to any single person in its totality. In short, science requires a social structure within which research takes place.

The idea that science is a spontaneous order has implications for our view of 'rationality'. We discuss two. First, as discussed already, the rationality of rational action must always be partially tacit. Second, rationality is a time-dimensioned process. Together, these two propositions say something about the nature of rationality in scientific enterprise and about the limits and possibilities of centralised direction of science.

Rationality is necessarily partially tacit as shown by our counting argument. If the counting argument is right and if scientists are rational, then part of what it means to be rational must remain outside any model of scientific activity. We may be 'rational' in ways we cannot say. The counting argument comes from Hayek's theory of cognitive psychology contained in *The Sensory Order* (Hayek 1952a). There, the central nervous system is modelled as a complex spontaneous order which functions as an apparatus for classifying and reclassifying nervous impulses. The classification of nervous impulses transforms them into sensations. The structure of the mind for Hayek is hierarchical; it is a relational ordering governed by the system of rules that classify nervous impulses. In this account, the mind is not 'governed by' a system of rules; it *is* a system of rules. A key point in Hayek's theory is the necessity of some (higher) mental activity to be governed by rules that cannot be articulated or made explicit. This follows from Hayek's claim that 'any apparatus of classification must possess a structure of a higher degree of complexity than is possessed by the objects which it classifies' (Hayek 1952a, p. 185); consequently, we can never fully explain our own thoughts and minds. If Hayek's theory is correct, each of us engages in mental activities not all of which we are able to articulate. That some of our knowledge must always be tacit is not an assumption, but an implication of Hayek's theory of complex phenomena applied to the mind. If Hayek is right, we can never fully step outside ourselves because we lack the Archimedean epistemological lever that would make it possible.

Rationality within a spontaneous order requires acknowledging that one's scientific knowledge is subject to important constraints. As mentioned above, there is a division of knowledge in science. But another constraint on knowledge exists when we consider science as a spontaneous order. As mentioned earlier, spontaneous orders are open-ended. They have no end-point. This is why spontaneous order theorists make use of evolutionary metaphors and models. Scientific knowledge in this sense has no purpose, no final objective apart from the particular aims of individual scientists. Perhaps the most sensible way of describing what scientific communities do, even though it may not be intentional, is the elimination of the relatively 'blacker' claims; science at its best is an error-correcting process. We use the vague adjective 'blacker' as a way to avoid the pretence of knowing precisely what it is that makes one claim, proposition, or theory preferable to its rivals. The appropriate standards for that are endogenously generated by the scientific process and often tacitly known by the scientific community.

This appropriately suggests understanding science itself as an evolutionary undertaking wherein some proposition is accepted as relatively correct only after other rival ones have been rejected as less true. But this involves neither the instantaneous acceptance of one proposition nor the immediate rejection of another. Instead, the acceptance of a theory (and the rejection of its rival) occurs in the context of a scientific tradition. The tension between established ideas and new rivals is not resolved with, say, one falsifying test, but through a potentially long scientific process of debate (Weimer 1979). It is in this sense that rationality in science is processive and not instantly specifiable. Rationality in science is not concerned with the content of knowledge, but with the kinds of activities consistent with the elimination of scientific error. The kind of activities we want will adhere to rules that give the widest play to critical discourse (Bartley 1984). Critical discourse is not just conversation. It is conversation subject to an evolved set of rules, which tend to promote honest criticism and frustrate immunising strategies.

If science is a complex spontaneous order for which only an explanation of the principle is possible, then the problem of reflexivity would seem to be much less acute than generally allowed. A theory is 'reflexive' if it may be applied to itself. Any scientific study of science must be reflexive. If one's theory of science does not apply to itself, there is a troublesome failure of reflexivity. Is the theorist given special immunity from the causal laws supposedly governing other types of scientists? The counting argument given above shows that any model of science will simplify. The model scientist is necessarily less complex than the real scientist. The theory does not and cannot claim to provide a complete causal explanation of the scientist's thoughts and actions. It can only identify some broad patterns of action and interaction. These broad patterns should generally apply to the scientific study of science as well. But if they do, the fact will have few or no implications. A judge may know that judges in civil trials tend to seek out the more reasonable set of expectations. He or she is not able to infer from that general knowledge which litigant should prevail. A firm manager may know that prices tend to equilibrium without feeling obliged to adjust his or her price toward its unknown, long-run equilibrium value. A causal explanation of science that gives only explanation of the principle is not likely to generate paradoxes and asymmetries when applied to itself.

## 4. CONCLUSION

We have argued that science is a spontaneous order. If we are more or less right, then we may have a partial explanation of the trend of 20th century philosophy of science toward increasingly complex representations of science. If science is a spontaneous order, then science is complex. If so, the relatively simple pictures of traditional philosophy of science were destined to be supplanted by accounts that are more complicated. These, in turn, were destined to be replaced by accounts that are even more complicated, and so on. In the last half-century, philosophies and theories of science have been giving increasingly complicated accounts of their object of study. If science is a spontaneous order, then science is a social process. In the last half-century, we have seen philosophies and theories of science give increasing attention to the social dimension of science. If science is a spontaneous order, then scientific rationality is processive. In the last half-century, we have seen philosophies and theories of science give subtle and less static accounts of scientific rationality. Our model of science was meant to fit science, but it seems to provide some insight into developments in science studies as well.

If we are right to view science as a spontaneous order, then we need not choose between modern and postmodern images of science as police report or detective novel. The suitable metaphors are not authored texts, but other complex evolving systems such as the legal system or the economy. The view that science is a spontaneous order gives full scope to several key points of postmodern views. We can recognise, for instance, that science is a social process in which scientists are not often animated by a pristine love of truth. But we need not infer from such insights that truth and reality are constructed. Truth is a regulative principle to which the overarching process conforms more or less satisfactorily in spite of the sometimes low motives of participants in that process. Selfish lawyers may get innocent clients off the hook. Selfish scientists may discover cures for deadly diseases. The economics of science will provide useful analyses if its practitioners do not try to claim greater powers for it than can be achieved in any discipline studying complex spontaneous orders.

## NOTES

1 For comments on earlier versions we thank Mie Augier, Peter Boettke, Paul David, Steven Fuller, David Levy, Maria Minniti, Larry Moss, Ivan Pongracic, Jr., Esther-Mirjam Sent, and the editors of this volume. Unfortunately, we must absolve all of them of blame for errors or infirmities of the text.
2 The phrase 'growth-of-knowledge' seems to have been coined by Lakatos and Musgrave who entitled their important edited volume *Criticism and the Growth of Knowledge*. Their use echoes the subtitle to Popper's 1963 book, *Conjectures and Refutations: The Growth of Scientific Knowledge*. Caldwell included the following footnote in the second edition of *Beyond Positivism*. 'I included too many writers under the "Growth-of-Knowledge" label, which only refers to the work of Lakatos and Popper. My mistake was to infer from the title of the Lakatos and Musgrave (1970) volume that all of its contributors (which included Kuhn and Feyerabend) were growth of knowledge theorists' (1994, p. xv). In our view, the genie is out of the bottle. The term 'growth of knowledge' has been applied to many thinkers beyond the tradition of Popper and Lakatos. In this chapter we use the term in Caldwell's earlier and broader sense.
3 'By 'order' we shall throughout describe a state of affairs in which a multiplicity of elements of various kinds are so related to each other that we may learn from our acquaintance with some spatial or temporal part of the whole to form correct expectations concerning the rest, or at least expectations which have a good chance of proving correct' (Hayek 1973, p. 36, emphases suppressed).
4 As far as we know, this distinction has not been made explicit in the complexity literature. The distinction may have been obscured by the implicit assumption that the performance of an adaptive order of the second kind can always be measured (in principle) by some pay-off function. For example, Holland makes this assumption (1992, pp. 20–31). But in the case of economies and ecological systems, the assumption may not be appropriate. There may be no 'performance measure' for the economy as a whole or the biosphere as a whole. But their elements do adapt. Brown identifies ecological systems as 'adaptive' because of three 'mechanisms' of adjustment. First, individual organisms adapt 'facultatively'. Plants spread their leaves out in the sun, for example. Second, whole species may migrate. Finally, natural selection works to adapt species to their circumstances (Brown 1994, pp. 422–3). These are three ways that the elements of an ecological system change to fit better in their environment. The ecological system as a whole, however, is not adjusting to its environment.
5 The other two insights are 'that what today is generally regarded as "social" or distributive justice has meaning only within the second of these kinds of order', that is, an organisation, and 'that the predominant model of liberal democratic institutions, in which the same representative body lays down the rules of just conduct and directs government, necessarily leads to a gradual transformation of the spontaneous order of a free society into a totalitarian system conducted in the service of some coalition of organized interests' (Hayek 1973, p. 2).
6 Most of this paragraph is cribbed from Koppl (1997).
7 We are purposely ignoring some logical difficulties of this claim. A spontaneous order produced by the interactions of many agents will be more complex than any of these agents only if the complexity of each agent is judged by its behaviour in relation to the overall order. In other senses, the agents may be as complex as one wishes.

## REFERENCES

Bacon, Francis (1955), *Selected Writings of Francis Bacon*, Hugh G. Dick (ed.), New York, NY: The Modern Library.
Bartley, William W. III (1984), *The Retreat to Commitment*, London: Open Court.

Bloor, David (1976), *Knowledge and Social Imagery*, London: Routledge & Kegan Paul.

Brown, James H. (1994), 'Complex Ecological Systems', in George A. Cowan, David Pines, and David Meltzer (eds), *Complexity: Metaphors, Models, and Reality*, Reading, MA: Addison-Wesley, pp. 419–450.

Bury, J. B. (1955), *The Idea of Progress: An Inquiry into its Growth and Origins*, New York, NY: Dover Publications.

Butos, W. N. and P. J. Boettke (2002), 'Kirznerian Entrepreneurship and the Economics of Science', *Journal des Economistes et des Etudes Humaines*, **12** (1): 119–130.

Butos, William N. and Thomas J. McQuade (2001), 'Mind, Markets, and Institutions: The Knowledge Problem in Hayek's Thought', in J. Birner, P. Garrouste and T. Aimar (eds), *F. A. v. Hayek As a Political Economist*, London: Routledge, pp. 113–33.

Caldwell, Bruce (1982), *Beyond Positivism: Economic Methodology in the Twentieth Century*, London: George Allen & Unwin.

Carnap, Rudolf (1958), *Introduction to Symbolic Logic and its Applications*, New York, NY: Dover.

Gray, John (1984), *Hayek on Liberty*, New York, NY: Basil Blackwell.

Hayek, Friedrich A. von (1952a), *The Sensory Order: An Inquiry Into the Foundations of Theoretical Psychology*, Chicago, IL: University of Chicago Press.

Hayek, Friedrich A. von (1952b), *The Counter Revolution of Science: Studies in the Abuse of Reason*, Chicago, IL: University of Chicago Press.

Hayek, Friedrich A. von (1967a), 'Degrees of Explanation', in Friedrich A. von Hayek (ed.), *Studies in Philosophy, Politics, and Economics,* Chicago, IL: University of Chicago Press, pp. 3–21.

Hayek, Friedrich A. von (1967b), 'The Theory of Complex Phenomena', in Friedrich A. von Hayek (ed.), *Studies in Philosophy, Politics, and Economics,* Chicago, IL: University of Chicago Press, pp. 22–42.

Hayek, Friedrich A. von (1967c), 'The Results of Human Action, but not of Human Design', in Friedrich A. von Hayek (ed.), *Studies in Philosophy, Politics, and Economics,* Chicago, IL: University of Chicago Press, pp. 96-–105.

Hayek, Friedrich A. von (1967d), 'Rules, Perception and Intelligibility', in Friedrich A. von Hayek (ed.), *Studies in Philosophy, Politics and Economics,* Chicago, IL: University of Chicago Press, pp. 43–65.

Hayek, Friedrich A. von (1973), *Law, Legislation and Liberty, Volume I: Rules and Order,* Chicago, IL: University of Chicago Press.

Holland, John H. (1992), *Adaptation in Natural and Artificial Systems: An Introductory Analysis with Applications to Biology, Control, and Artificial Intelligence*, Cambridge, MA: MIT Press.

Kauffman, Stuart A. (1993), *The Origins of Order: Self-Organization and Selection in Evolution*, New York, NY and Oxford: Oxford University Press.

Keynes, John M. ([1921] 1973), *A Treatise On Probability*, reprinted as Volume VIII of D. E. Moggridge (ed.), *The Collected Writings of John Maynard Keynes*, New York, NY: St. Martin's Press.

Kitcher, Philip (1993), *The Advancement of Science*, New York, NY and Oxford: Oxford University Press.

Koppl, R. (1992), 'Invisible-Hand Explanations and Neoclassical Economics: Toward a Post Marginalist Economics', *Journal of Institutional and Theoretical Economics,* **148** (2): 292–313.

Koppl, Roger (1994), 'Invisible-Hand Explanations', in Peter J. Boettke (ed.), *The Elgar Companion to Austrian Economics*, Cheltenham: Edward Elgar, pp. 192–6.

Koppl, Roger (1997) 'Review of Uskali Mäki, Bo Gustafsson, and Christian Knudsen (eds.), *Rationality, Institutions and Economic Methodology'*, in *Advances in Austrian Economics,* **4**: 241–5.

Koppl, Roger (2000a), 'Policy Implications of Complexity: An Austrian Perspective', in David Colander (ed.), *The Complexity Vision and the Teaching of Economics*, Cheltenham: Edward Elgar, pp. 97–117.

Koppl, Roger (2000b), 'Teaching Complexity: An Austrian Approach', in David Colander (ed.), *The Complexity Vision and the Teaching of Economics*, Cheltenham: Edward Elgar, pp. 137–146.

Kuhn, Thomas S. (1970), *The Structure of Scientific Revolutions*, Second edition, enlarged, Chicago, IL: University of Chicago Press.

Lakatos, Imre and Alan Musgrave (eds) (1970), *Criticism and the Growth of Knowledge*, Cambridge: Cambridge University Press.

Langlois, R. N. (1989), 'What Was Wrong With the "Old" Institutional Economics? (And What Is Still Wrong With the "New"?)', *Review of Political Economy*, **1** (3): 272–300.

Latour, Bruno (1987), *Science in Action: How to Follow Scientists and Engineers Through Society*, Cambridge, MA: Harvard University Press.

Latour, Bruno and Steve Woolgar (1979), *Laboratory Life*. London: Sage.

Lavoie, Don (1985), *National Economic Planning: What Is Left?,* Cambridge, MA: Ballinger.

Machlup, Fritz (1936), 'Why Bother With Methodology?', *Economica*, **3** (9): 39–45.

Merton, Robert K. (1957), 'Science and the Social Order', in Robert K. Merton (ed.), *Social Theory and Social Structure*, revised and enlarged edition, New York, NY: The Free Press, pp. 537–49.

McQuade, Thomas J. and William N. Butos (2002), 'Order-Dependent Knowledge and the Economics of Science', mimeo.

Neurath, Otto ([1938] 1955), 'Unified Science as Encyclopedic Integration', in Otto Neurath, Rudolf Carnap and Charles Morris (eds), *International Encyclopedia of Unified Science*, Volume 1, Nos. 1–5, Chicago, IL: University of Chicago Press, pp. 1-27.

Pickering, Andrew (1992), 'Introduction – From Science as Knowledge to Science as Practice', in Andrew Pickering (ed.), *Science as Practice and Culture*, Chicago, IL: University of Chicago Press, pp. 1-28.

Polanyi, M. (1962), 'The Republic of Science: Its Political and Economic Theory', *Minerva,* **1**: 54–73.

Popper, Karl R. (1963), *Conjectures and Refutations: The Growth of Scientific Knowledge*, London: Routledge.

Schutz, Alfred ([1932] 1967), *The Phenomenology of the Social World*, Evanston, IL: Northwestern University Press.

Suppe, Frederick (1977), 'Afterword', in Frederick Suppe (ed.), *The Structure of Scientific Theories*, Second edition, Urbana, IL: University of Illinois Press, pp. 615–728.

Weimer, Walter B. (1979), *Notes on the Methodology of Scientific Research*, Hillsdale, MI: Lawrence Erblaum.

Wible, James R. (1998), *The Economics of Science: Methodology and Epistemology as if Economics Really Mattered*, London: Routledge.

# 9. Must spontaneous order be unintended? Exploring the possibilities for consciously enhancing creative discovery and imaginative problem-solving

**Mathew Forstater**

The conventional view has it that spontaneous order is unintended. Drawing on Hayek's distinction between *cosmos* and *taxis* – and Michael Polanyi's parallel distinction between spontaneous order and corporate order – most scholars identify *cosmos* or spontaneous order as the result of 'human action but not of human design', and *taxis* or corporate order as the result of intentional design or planning. Some have explored combinations of the two, a number of examples of which have been offered by Allen (1998, pp. 188ff). There are arrangements that are *taxes* in their inception – deliberately established futures markets, for example – but that are allowed to operate in many respects spontaneously once the rules have been laid down. Other *taxes*, such as some unsupervised work groups, are organised to meet specified goals, but operate without command and are permitted to develop based on 'spontaneous mutual adjustments' (Allen 1998, p. 189). Other corporate associations – many households and social clubs fit this category – are not conceived to meet (though they themselves may set) specific goals, but also mutually adjust in a spontaneous manner. There are also *taxes* within the *cosmos* – Coase long ago characterised the firm within the larger market as such.

But Allen describes Hayek's 'Great Society' as 'a *cosmos* without any element of *taxis*' (Allen 1998, p. 187), and planning is depicted as the polar opposite – pure command. Moreover, Allen characterises Polanyi's notion of science as a *cosmos* model for a 'free society', though he notes that there are aspects of *taxis* in Polanyi's vision of a 'republic of science' (Allen 1998, pp. 190–1). A crucial task in this type of approach is thus focused on finding the right mix of *cosmos* and *taxis*, both in terms of the proportions of each and the character of the constituent parts.

My point of departure is the position that virtually the whole conceptual framework regarding – and epistemological approach to – the discussion of 'orders' is fraught with problems. At the root is the dichotomous and oppositional treatment of dualistic categories – *taxes* vs. *cosmos*, intuition vs. logic, market vs. state, spontaneity vs. design, tacit vs. explicit knowledge, and so on. Rather than trying to replace the terminology, I want to attempt to treat the categories differently. If successful, there will be greater fluidity among the oppositionals and the dichotomous dualisms would instead appear as interrelated aspects. All this is important because the standard treatment has somewhat severe implications. For example, Allen's treatment overstates the order-bestowing power of the price system and its associated incentive structures; under-emphasises the impact of changing historical contexts; considers the desire for social justice – and even 'face-to-face relationships, solidarity, and working together for common purposes' – as deriving from a nostalgia for pre-capitalist communities; and more (Allen 1998, pp. 155–6; 185–7).

I want, then, to focus on a slightly different interplay of *taxis* and *cosmos*. First, I want to argue that it is not simply a matter of a 'mix' of differing orders. Such an exercise, while useful and perhaps even necessary, still treats the different orders in a dichotomous manner. Rather, I want to argue that, on the one hand, there are intentional aspects to the processes associated with spontaneous order, that are difficult to separate from the unintended or spontaneous aspects of *cosmos*, and the relation is not simply 'additive'. On the other hand, I want to argue that spontaneity can, does, and should play a role in *taxes*. Related to this are Polanyi's notions of tacit and explicit knowledge. I want to argue that both tacit and explicit knowledge are vital to both spontaneous and designed orders. Furthermore, as Polanyi points out, tacit knowledge can become explicit, and this is important for understanding the interplay of intended and unintended aspects of spontaneous and designed orders.

Crucial to both spontaneous and designed orders are creative discovery and problem-solving, and I want to explore the possibilities for the conscious enhancing of both. Are there methods and strategies for enhancing the powers of creative discovery and problem-solving? I think there are and elsewhere (Forstater 1999) I have catalogued some of them, drawing on the work of Polanyi, the mathematician George Polya, and also pragmaticist philosopher Charles Sanders Peirce. Both Polya's heuristics and Peirce's abduction or retroduction are methods of problem-solving and creative discovery, and both are related to the method of working backwards. Other notions with family resemblances to some of the ideas of all these traditions – Schutz's common-sense, C. Wright Mills's sociological imagination, Lowe's instrumental analysis and spontaneous conformity, Vygotsky's free play, and others will also be considered.

Dewey (1916) said that experience is 'pregnant with connexions', that experience is 'full of inference' and 'experience in its vital form is experimental, an effort to change the given'. I want to explore inference in science (that is, *inquiry*), inference in everyday life, inference in participatory democratic associations, inference in planning and policy, inference in education. In his *Democracy and Education* Dewey wrote that 'A democracy is more than a form of government; it is primarily a form of associated life, of conjoint communicated experience' (Dewey 1916, p. 101). I want to explore 'the method of free and effective intelligence' in all human experience and association, and the emancipatory potential of inquiry. I also want to explore the subtle interplay of spontaneity and intentionality in both social inquiry and social life, and in spontaneous and designed associations.

In his discussion of 'Genius in Science', Polanyi explicitly noted the outwardly contradictory interplay of intentional and spontaneous elements in scientific discovery:

> Genius is known for two faculties which may seem incompatible. Genius is a gift of inspiration, poets back to Homer have asked their muse for inspiration, and scientists back to Archimedes have acknowledged the coming of a bright idea as an event that suddenly visited them. But we have ample evidence of an opposite kind; genius has been said to consist in taking infinite pains, and all kinds of creative pursuits are in fact extremely strenuous. How can these two aspects of genius hang together? Is there any hard work, which will induce an inspiration to visit us? How can we possibly conjure up an inspiration without even knowing from what corner it may come to us? And since it is ourselves who shall eventually produce the inspiration, how can it come to us as a surprise? Yet this is what our creative work actually does. It is precisely what scientific discovery does: We make a discovery and yet it comes as a surprise to us. The first task of a theory of creativity, and of scientific discovery in particular, must be to resolve this paradox. (Polanyi [1972] 1997, p. 268)

The key to the solution, as Polanyi had earlier argued elsewhere ([1966] 1997), is to be found in ordinary perception, and in fact he argued that science is an extension of perception, a Gestalt-like 'integration of parts to wholes' that is intentional and yet 'in an important sense' spontaneous; that is experienced as an inspiration, itself 'evoked by the labours of the thrusting imagination' (Polanyi [1972] 1997, pp. 268–9, 275). The ability to perceive the coherence of the working whole, the organism as it were, and the joint functioning of its parts begins with the identification of a problem and the conscious decision to pursue it. For Polanyi, this entails personal judgment as to the importance of the problem and one's ability to solve it, which he considers to constitute a guess. Polanyi's view, as he often noted, bears close resemblance to that of Polya, who has done much to outline the strategies for creative discovery and imaginative problem-solving (Forstater 1999). It also has important connections to Peirce's abduction or retroduction, which, unlike deduction or induction, is capable of producing new knowledge.

Retroductive inference or heuristics seems beset by the contradiction: on the one hand, word such as 'guess', 'instinct' and 'imagination' are invoked to discuss the process, while on the other hand it is insisted that there is a 'logic' of scientific discovery. Some light may be shed on this topic, as well as the difficulty of dealing with these issues in abstract terms, through Michael Polanyi's distinction between two different types of knowledge (1958; 1959; 1966). By 'explicit knowledge' Polanyi refers to knowledge that is articulate, that which is usually intended by the word 'knowledge', that is written words, mathematical formulae, maps, and so forth (1959, p. 12). But Polanyi identifies 'tacit knowledge' as the 'dominant principle of all knowledge' which 'at all mental levels ... [is] decisive' (1959, pp. 13, 19). Tacit knowledge is 'unformulated'; it is the 'knowledge we have of something we are in the act of doing' (1959, p. 12). Explicit knowledge can be critically reflected upon, which is an advantage that it has over tacit knowledge (1959, pp. 15–18). Yet tacit knowledge concerns discovery, which is the basis for explicit knowledge. As Polanyi puts it, a traveller with a detailed map enjoys superiority over the explorer who first enters a new region: 'yet the explorer's fumbling progress is a much finer achievement than the well-briefed traveller's journey' (1959, p. 18). Or, put another way: 'Even if we admitted that an exact knowledge of the universe was our supreme mental possession, it would still follow that our most distinguished act of thought consists in *producing* such knowledge' (1959, p. 18). Key, also, for present purposes, is that Polanyi argued that methods can be devised by which ways of knowing that are tacit can be made explicit.

Because of its nature, the 'way of discovery' (Gelwick, 1977) is difficult to explicate. Polya has thus identified the 'first task' as that of 'collect[ing] and classify[ing] such problem-solving procedures' and to 'develop a repertory of problem-solving techniques' (Polya [1971] 1984, p. 590). Even so, this will not solve the issue comprehensively, because there remains the issue of choosing from among the available techniques, a decision which will require that the investigator 'use personal judgement, as Polanyi would say' (ibid.). This is similar to Adolph Lowe's discussion of choosing among alternative hypotheses: 'there are no binding rules, according to which a researcher could decide in favour of one among many possible hypotheses. Which one he chooses in the end, adopting ... Einstein's "free creation of the mind", is neither a strictly determinable nor an arbitrary decision' (Lowe 1992, p. 327).

Polya and Polanyi have both contributed to the challenge of explicating the procedures of the inexplicable. Whereas Polya's efforts have been more along the lines of taking an inventory of tools, Polanyi has explored the tacit fringes of these procedures. For Polanyi, appreciation of a problem is itself part of the act of discovery (Polanyi 1958, p. 121). Seeing a problem 'is a definite addition to our knowledge', and 'to recognise a problem that can be solved and worth solving is a discovery in its own right' (1958, p. 120). In the process of grappling with a problem, a 'heuristic stress' builds, which is akin to an emotional strain on the part of the investigator. Discovery leads to a release, such as running through the streets crying 'Eureka!' (1958, p. 122).

One heuristic tactic noted by Polanyi is to continuously reorganise the problem 'with a view to eliciting some new suggestive aspects of it' (Polanyi 1958, p. 128). This is reminiscent of C. Wright Mills's suggestion that 'the re-arranging of the [researcher's] file ... is one way to invite the [sociological] imagination' (Mills 1959, p. 212):

> Imagination is often successfully invited by putting together hitherto isolated items, by finding unsuspected connections ... As you rearrange a filing system, you often find that you are, as it were, loosening your imagination. Apparently this occurs by means of your attempt to combine various ideas and notes on different topics. It is a sort of logic of combination, and 'chance' sometimes plays a curiously large part in it. In a relaxed way, you try to engage your intellectual resources ... Of course, you will have in mind the several problems on which you are actively working, but you will also try to be passively receptive to unforeseen and unplanned linkages.

Both Polanyi and Mills relate this 'reorganising' tactic to another, what Polanyi refers to as 'ransack[ing] our memory for any similar problem' (Polanyi 1958, p. 128) and Mills calls 'get[ting] a *comparative* grip on the materials' (Mills 1959, p. 215, original emphasis). This is actually what Polya refers to in his story about a person trying to cross a creek when he states that 'the man may recall he has crossed some other creek by walking across a fallen tree' (Polya [1945] 1957, p. 145) and also what Hobbes points to when he writes that 'from desire ariseth the thought of *some means we have seen produce the like of that which we aim at*' (Polya [1952] 1981, p. 2, emphasis added), in other words, we are familiar with an analogous problem that has been solved: 'Any conjecture, of course, must have been suggested ... by somehow related ideas (special cases, analogies, etc.), although, perhaps, at the moment of conceiving the conjecture those ideas were not clearly and explicitly present' (Polya [1948] 1984, p. 474). Polya distinguishes 'similarity' from 'analogy' as two related but distinct heuristic tools.

In the course of the heuristic search, we must look for 'favourable signs', which of course must not be mistaken for 'proof', but which encourage 'further investigation' (Polya [1949] 1984, p. 490). Lowe as well cautions that 'the findings of heuristic analysis can be accepted only provisionally' (Lowe 1992, p. 327). Polya invokes the notions of the 'bright idea' and 'feeling we are "on the right track"' to get at the seemingly intuitive aspects of the discovery procedure. For Polanyi (1958, p. 128), 'success depends ultimately on the capacity for sensing the presence of yet unrevealed logical relations between conditions of the problem, the theorems known ... and the unknown solution...'. Polanyi invokes the 'common experience(s) of groping for a forgotten name' and searching for a name or word that is said to be 'on the tip of the tongue' to illustrate the 'sense of growing proximity to the solution' that guides discovery (Polanyi 1958, pp. 128–9). As Lowe puts it, the 'researcher "senses" a structural relationship between the hypothesis he chooses and the problem he wants to solve' (Lowe 1992, p. 327).

Equally important is Polanyi's suggestion that self-awareness of the capacity to sense the 'accessibility of a hidden inference', as well as of the ability to 'invent transformations of the premisses which would increase accessibility' is a 'foreknowledge' which itself 'biases our guesses in the right direction' (Polanyi 1958, p. 129). The discovery-enhancing effects of our awareness of our ability to discover is also related by Polanyi to the fact that 'a set purpose may automatically result in action later on' as when we go to bed resolved to wake up at a certain hour, and then actually do. These factors also help explain the 'self-accelerating manner of final stages of solution', in other words, the closer we get the faster we progress. These aspects of discovery are not treated lightly by Polanyi, who takes the position that 'the whole process of discovery and confirmation ultimately relies on our own crediting of our own vision' (Polanyi 1958, p. 130).

Peirce also believed that abductive reasoning was 'a skill that could be improved by practice or discipline' (Ochs 1993, p. 61). Like Polya, Polanyi, and Mills, Peirce sees a vital role for 'common-sense', a view that has points of contact with Schutz as well (Schutz [1953] 1967). To this must be added the value of imagination in making discoveries.

It must be emphasised again that all the authors referred to here are of the opinion that these processes are complementary to the generally recognised procedures of scientific practice. But the point is that these processes are crucial and indispensable, and recognition of this increases their power.

What unites the detective, the physician, the scientist, the artist? The quest, the discovery process, following clues, imaginative guesswork, seeking to solve the mystery, to find the coherence among the parts, searching for solutions – by any means necessary. Polya reports solving proofs through dreams, and he once identified a mathematical problem and a found a clue to its solution when he by 'chance' encountered twice in one afternoon an acquaintance and his friend on a series of winding, connected footpaths in a park. Are there ways of consciously increasing one's ability to 'guess' correctly? Indeed, among the kit of tools of discovery are included means of increasing one's 'luck'.

Austin (1978) has identified different 'personality traits' associated with various kinds of luck or chance, and that thereby 'facilitate discovery and invention' (Accardo 1987, p. 67). Hotson has put forward a similar argument with respect to serendipity, and while the particular application is with regard to literary research, the strategies for serendipity-enhancement may be applied in other areas as well. Hotson reports that Walpole coined the term 'serendipity' after hearing a story about the 'Three Princes of Serendip', in which the Royal Trio continuously made favourable discoveries by accident as they travelled around. Hotson, like Austin, identifies curiosity as one of the traits inviting good fortune. Quoting Simon Flexner:

> Curiosity, not utility, is the master key to human knowledge; curiosity which may or may not result in something useful. And the less that curiosity is asked to justify itself day by day, the more likely it is not only to contribute to human welfare, but to the equally important satisfaction of the human mind. (Hotson 1942: 80)

Echoing Accardo's summary of Austin to the effect that 'chance favors those in motion' and the 'Kettering Principle: I have never seen anyone stumble onto something sitting down', Hotson also emphasises that:

> When all is said, the essential to bear in mind about serendipity – whether you call it happy accident or lucky chance – is that the Princes had to travel; and travel means labor. The searcher has to go into them thar hills, and then look about him and dig at twenty to the dozen. You don't strike devilish good luck without weevilish hard work. (Hotson 1942: 81)

Persistence is likewise mentioned by both Austin and Hotson, the latter ending by quoting Shakespeare to the effect that: 'Tomorrow I intend to hunt again' (Hotson 1942: 94), determined to follow the right 'clues, out of the mass … most likely to lead to the lucky spot' (ibid.: 87). Once more, or many times over, we return to the delicate tapestry of intention and spontaneity, chance and design.

Elsewhere (Forstater 1999), I have argued that the strategies and tactics for the enhancement of the powers of creative discovery described herein might also be employed in the sphere of economic policy; that Adolph Lowe's instrumental analysis may best be thought of as a kind of policy discovery procedure, and that this opens up the way for another kind of potential hermeneutic Austrian contribution to public policy. I also argued that an overly dichotomous treatment of market vs. state in some Austrian writings constitutes an obstacle to realising this potential. And this chapter started out by challenging the related overly dualistic or dichotomous treatment of spontaneous vs. designed orders, intentional vs. spontaneous processes within both *cosmos* and *taxes*, and, more generally, reason vs. imagination, intuition vs. inference, and so on. Instead, we are arguing for a more subtle and combinatory treatment. Continuing along these lines, I want now to present another concept of Adolph Lowe's – that of 'spontaneous conformity' – along with his and others' related thoughts on freedom, education, and socialisation, that get to some of the other issues at stake in the discussion of spontaneous orders and related ideas, which in turn brings us back to the issues of social and economic life with which most of us are concerned.

Lowe was a colleague of Michael Polanyi's at Manchester in the 1930s, where they were engaged in discussions around these and related themes, and during which time Polanyi first developed his concept of spontaneous order and Lowe first developed his idea of spontaneous conformity (Lowe 1937a; 1937b; 1940). Lowe returned again and again throughout the course of his life to what he referred to as the great 'riddle': 'how is freedom of [individual] choices compatible with integral [social] order?' (Lowe 1942: 445). Clearly, the problem of freedom and order is ultimately what the whole 'spontaneous order' theme is all about. Given the behavioural requirements necessary for determinate economic outcomes, Lowe (1935, p. 62; 1951: 413) notes that the only alternative to the authority of a command system is voluntary restriction of goal-inadequate behaviours and the obeying of a general code of conduct. If determinate behaviour is not just any behaviour (in other words, if what is sometimes called 'licence' is unlikely to be compatible with socioeconomic order) and if determinate behaviour is not to be imposed from without, then such a code must be internalised, and ideally 'experienced as ... spontaneous decision' (Lowe 1942: 439–40). Lowe refers to this process as 'spontaneous conformity', which Clary (1998, p. 276) has defined as 'the spontaneous consensus among the members of society to a social code of conduct, the standards of which are accepted and obeyed by the individual members of society'. Such self-restriction is 'the price of political and economic freedom' (Lowe 1935, p. 71; cf. 1937a and 1942: 440).

In fleshing out these issues, Lowe identifies '*education* in the widest sense of the term' as crucial to the socialisation processes necessary to reconcile freedom and order in liberal society (1988, p. 128, original emphasis). Lowe refers here not merely to intellectual training or information necessary for comprehending the social implications of individual behaviour in the more technical sense. Rather, he associates successful education in this broad sense with 'integrating the rational with the moral' and '*commitment to a life-ordering principle*' (1988, p. 130, original emphasis).

Like his writings on economic theory, history, methodology, policy, and political philosophy, Lowe's writings on education span his career. What is probably his least-known book, *The Universities in Transformation*, was published in 1940, but already Lowe had anticipated that work with an article, 'The Task of Democratic Education', comparing university education in pre-Hitler Germany and England (1937b), as well as a related discussion of education in *The Price of Liberty*, published the same year. Lowe spoke on the topic of education often between the 1940s and 1970, when he addressed the question 'Is Present-Day Higher Learning Relevant?' both at Columbia University's 'Seminar On the Nature of Man' and the General Seminar at the Graduate Faculty of Political and Social Science at the New School for Social Research, subsequently publishing the paper in *Social Research* in 1971. Education is a central theme once more in Lowe's last book, *Has Freedom a Future?* (1988).

Writing after four years in exile and with World War II imminent, Lowe makes clear in 'The Task of Democratic Education' that it will no longer be adequate merely to profess the traditional postulates of democratic education: 'intellectual freedom and personal responsibility. We have to *prove* that these liberal principles ... are superior to the new gospel of indoctrination and enforced conformity' (1937b: 381, emphasis added).

Lowe then immediately turns to an issue which forms a crucial part of his life-work, the social basis for individuation. Far from denying the importance of individual freedom and personality, he refers to their role as serving a 'leavening function, working on a dough which is composed of very different material ..., the pre-liberal heritage of attitudes and life patterns'. For Lowe (1937b: 382), the disintegration of this communal base threatens the survival of free society:

> If this proves true, the task of democratic education extends far beyond the cultivation of freedom and personality. For the preservation of these liberal values modern democracy will have to undertake a much bigger task: that of reviving, or even creating, the substance of a new social and economic order.

Lowe goes on to compare the pre-Hitler German university system, characterised by a very high degree of academic freedom and independent research, with the English system, which Lowe views as 'rearing grounds for a social type' (1937b: 385). In the first half of the 19th century, 60 per cent of the students in Germany were the children (sons) of civil servants, teachers and clergy, which meant that socialisation was the province of the 'feudal and military standards of the Prussian tradition' (ibid.: 383). But this 'division of labour' continued right up to World War I, by which time 50 per cent of the students were now from the business classes.

The German system produced some excellent scholars, but the majority of students, Lowe laments, were not up to the challenge of self-guided education, and submitted to a 'dull, though well-organised, cramming system' (1937b: 383). Such a system was unable to shape and encourage a social consciousness in the students, who received their real education – social and physical –

> ... in their fraternities and student corps. There, however, not the humanistic ideal of free thinking, but the Wilhelmian parody of Prussian tradition was instilled in them ... This dualism was much deplored as a sin against the true spirit of the German University. But as this university refused to do anything but train scholars and satisfy the desire for personal self-refinement, some other body had to step in to give the student masses human guidance and social drill. It was on this residue of the feudal and absolutist ages on which the leaders of the nation actually lived. (Lowe 1937b: 383–4)

The universities became overcrowded following World War I, and the unemployment of the inter-war period rose, with 50 000–70 000 unemployed graduates up to 1932 forming the basis of the 'propagandists and the officers of the counter-revolution' who were to turn 'an economic disaster into a general social upheaval' (Lowe 1937b: 384).

For Lowe, the German university failed at mass education, the fundamental challenge of modern democracy. England's success, on the other hand, Lowe attributes to its ability to meet the needs of the average student and 'produce a social character' (ibid.: 385). Though English universities experienced similar demographic trends to those in Germany, the English system fulfilled its function of shaping 'a general character pattern through the daily experience of a group life' (ibid.: 385).

As Lowe reiterates in *The Universities in Transformation*, from early on in life the socialisation of the individual is intimately linked to societal institutions, in particular educational institutions. This is as true in a totalitarian society as in a democratic one. But in the latter, the socialising forces must forge not a numbing, mindless uniformity; while still involved in the production of a 'definite human *type*', social institutions such as the educational system must be 'flexible enough to enable, and even encourage, the type to develop "beyond the type"' (Lowe 1940, p. 2, emphasis added).

For Lowe, the socialised individual of a democratic society must be 'dynamic', and he points to two crucial aspects of such a view: 'It recognises ... the "incompleteness" of human history, and points to the significance of cultural evolution .... [and] it by no means involves the uncritical acceptance

... of the actual ideas and standards prevailing at any given moment' (Lowe 1940, p. 2).

As such, Lowe points out that while England produced no great 'intellectual rebels against tradition' such as Marx, Nietzsche, and Freud, it produced 'the Friends, the Radicals, and the Fabians, that is, an intelligentsia whose members used the ruling social code for the most daring offensive against what they regarded as abuses of the true meaning of tradition' (1937b: 386–7). For Lowe, the characterisation of 19th century England as 'a liberal society' applies as a description of the economic system, but it must not be permitted to disguise the importance of voluntary associations and the underlying social fabric, which would not properly be described as 'atomistic' (1940, p. 6).

Lowe argued that the success of the English university system lay partly in the fact that its task was not as great: for Lowe, the English university had only to develop 'existing attitudes which are pre-formed by family tradition and daily experience [and] permeate the social conduct of the whole nation' (1937b: 386). Whereas in Germany: the humanistic education and the Prussian socialisation were socially and substantively separate, in England Lowe found an amalgamation of pre-liberal tradition and liberalism into 'one homogeneous life pattern: spontaneous conformity' (ibid.: 386).

This idea of 'spontaneous conformity' is a central thread in Lowe's work from the 1930s through the remainder of his life. His exile in England provided him the opportunity to observe English society in a detached manner, and he quickly became impressed with the apparent contradiction between individual freedom on the one hand and fairly strict social conformity on the other. Lowe's *The Price of Liberty* is dedicated to this phenomenon, and his conclusions had a significant influence on his writings from that time onward, regaining a central place in his last major work, *Has Freedom a Future*?[1]

Such a socialisation process instilling conformity in individuals may be thought of as the antithesis of 'spontaneity' or 'freedom'. More recent evidence regarding the socialisation process in young children demonstrates that far from being the result of strict and rigid 'training', the acquiring of social norms is a rather spontaneous process and the 'natural' outgrowth of free play and the imagination. Important contributions in this area may be found in the work of Vygotsky ([1934] 1962), an early 20th century psychologist who theorised about the way in which conceptual development occurs in children based on social practices.[2] Vygotsky's ideas may provide leads for better understanding Lowe's seemingly paradoxical notion of 'spontaneous conformity', as well as extending the focus on education to a variety of levels and areas.

To the outside observer, 'free play' (as opposed to organised, structured activities) appears as 'free' and 'spontaneous'. But, as Berk (1994: 33) notes in her overview of Vygotsky's ideas, 'free play is not really "free"': '[I]nstead, it requires self-restraint – willingly following social rules ... By enacting rules in make-believe, children come to better understand social norms and expectations and strive to behave in ways that uphold them'. For Vygotsky, a fundamental aspect of all imaginative or representational play is that it 'contains rules for behaviour that children must follow to successfully act out the play scene' (Berk 1994: 31, emphasis deleted). Thus, free play 'supports the emergence of the ... capacity to renounce impulsive action in favour of deliberate, self-regulatory activity' and so has a crucial role in development and socialisation (ibid.: 32). Dramatic and imaginative play therefore prepares young children for the more formal games with overt rules of middle childhood, 'which provides additional instruction in setting goals, regulating one's behavior in pursuit of those goals, and subordinating action to rules rather than to impulse – in short, for becoming a cooperative and productive member of society' (ibid.: 33). Far from being an idiosyncratic or utopian notion of Lowe's, 'spontaneous conformity' may be seen as being at the foundation of human socialisation and societal functioning. The work of Vygotsky and his followers demonstrates that there is no need for strict enforcement of conformity, which would be authoritarian and dehumanising. Quite the contrary: socialisation occurs naturally from early on in life as the outgrowth of imaginative play activities. In the context of Lowe's political economics, however, a question immediately arises. Vygotsky's theory describes the *process* of socialisation, of rule-following behaviour, but not the *content* of the rules themselves. The socialisation that takes place from early childhood in a given society reflects the already-existing social codes of conduct in that society. But at the core of Lowe's thesis is the idea that social codes are not universal and timeless, instead they are historically contingent and context-dependent. Historical social, technological, and environmental transformations alter the efficacy of what Lowe's colleague at the New School, the phenomenological sociologist Alfred Schutz ([1943] 1970), called social 'recipes'.[3] On the one hand, if social codes become rigidly fixed, they may be continuously passed on as tradition from generation to generation even after socioeconomic change has diminished their goal-adequacy. On the other hand, negative anti-social and otherwise goal-inadequate habits will also be imitated and adopted by children in socialisation processes.

One of the tasks of Lowe's instrumental analysis is to discover the suitable behaviours for setting the system on a goal-adequate path (Lowe [1965] 1977). A variety of methods may be employed in order to try to induce goal-adequate behaviour, ranging over context-making, moral suasion, public education, and 'enforcement' of varying degrees of severity and formality. Often, goal-attainment may require that rigidly ingrained habits be altered or broken, and replaced by new practices.

As Lowe emphasises, external controls will be unnecessary to the extent that self-control is employed: voluntary adherence to a social code will eliminate the need for direct government control. Eventually, new practices will be established which will then become part of the socialisation process. While there will be 'violations', the viability threshold will not be crossed.

The key, then, becomes the ability of society's members to alter their practices when necessary. Political economics thus requires a theory of self-control, which itself entails *habit-change*. We have already seen above that Lowe considers 'education in the widest sense' as key to self-control. Thus, the social function of education must not be limited merely to the process by which society's members become socialised to fixed behavioural codes: *the very ability to alter one's own habits, that is, self-control, and the skill of adapting one's behaviour in the face of changing circumstances and in the light of social necessities, must itself be part of the content.*

Peirce explicitly addressed these issues, which are not unrelated to his theory of abduction, a concept Lowe believed to be closely related to his own 'instrumental inference'.[4]

Peirce invented the term 'abductive reasoning' to refer to the 'inquiry we undertake to generate hypotheses about *how we might reform what we already do*. He believed this mode of reasoning was a power as well as a skill that could be improved by practice and by discipline' (Ochs 1993, p. 61, emphasis added). Peirce believed that in practice we use 'guiding principles', habits that, not unlike Schutz's social recipes, 'structure our behavior and experience' (Neville 1992, p. 29). Peirce was at first interested in the process of habit-formation, but soon after became interested in the question of what could be done if our habits prove faulty (Ochs 1993, p. 68). Peirce's investigations (1931–35, pp. 478–81) identified a number of sources of habit-change, the most important of which he considered to be 'efforts of the imagination':

> [G]iven free play, the imagination gives uninhibited expression to the fundamental categories of our existence in the contemplation of which inquirers *may construct norms for reforming our habits of action*. The product of abduction is of practical import, because it offers possibilities that might really be enacted within our contexts of action: possibilities of real habit-change. (Ochs 1993, pp. 71–2, emphasis added)

Thus the role of imagination in the creative construction and reconstruction (in the face of changing circumstances) of social norms is in no way limited to young children. Such a process continues into adulthood, and becomes increasingly self-determined. In addition, this is increasingly so in modern society. As Lowe (1988, p. 2) put it: 'From now on, the future will have to be more and more the result of our deliberate choices, at every level of human activity'.[5]

For Lowe, voluntary adherence to social codes of conduct eliminates the need for externally imposed controls. The family resemblances of Lowe's codes of conduct to Peirce's 'guiding principles' and Schutz's social 'recipes' have already been noted. It may be useful, then, to briefly examine Gorman's sharp critique of Schutz's notion of freedom, since it may also serve as an implicit critique of Lowe's similar conception of freedom associated with spontaneous conformity.

According to Gorman (1977, pp. 71–2), if social recipes are what determine behaviour:

> ... then there is little more than hypocrisy in contending that we are free, self-determining, meaning-endowing actors. [In this conception] the action we freely choose to perform is identical to the behaviour we would exhibit if this were impersonally determined by social typifications ... In the common-sense world, I am free only to obey.

If codes of conduct are prescribed by society, and we are socialised to obey the codes of conduct, then how is our choice to obey them 'freedom?' Gorman claims that in the Schutzian framework it is 'freedom' simply because that is how it has been defined; freedom here is *defined* as voluntarily obeying the rules.

Gorman's analysis cannot be easily dismissed. But Schutz is susceptible to the critique because he claims that individuals *will* choose to act in a way that coincides with objective necessity. Such an imposed coincidence of objective and subjective necessity is not to be found in Lowe, however. For Lowe, individuals must *decide* whether or not to follow the social rules. In fact, continuous critical evaluation of our habits is necessary in order to prevent perpetuating those which are no longer workable or desirable. Because Schutz does not adequately consider structural change and the impact such change has on the efficacy of social recipes, there is little space given to (or necessity for) critical self-consciousness.[6] But in a dynamic, transformational context, such critical self-consciousness is central to the adaptations required for societal functioning.

Gorman's depiction (1977, p. 83) of the Schutzian actor as 'naive [and] unquestioning ..., automatically responding to internalised social dictates' is somewhat reminiscent of C. Wright Mills's 'cheerful robot' (Mills 1959, p. 171). Mills believed that in what he called the 'post-modern period',[7] the individual suffers from Mannheim's 'self- rationalisation', conforming to the rules and regulations of the rational 'alienating organisation' (Mills 1959, pp.

166ff). Under such rationalisation, 'the guiding principles of conduct are alien to and in contradiction with all that has been historically understood as individuality':

> The society in which this man, this cheerful robot, flourishes is the antithesis of the free society – or in the literal and plain meaning of the word, of a democratic society. The advent of this man points to freedom as trouble, as issue, and – let us hope – as problem for social scientists. Put as a trouble of the individual – of the terms and values of which he is uneasily unaware – it is the trouble called 'alienation.' As an issue for publics – to the terms and values of which they are mainly indifferent – it is no less than the issue of democratic society, as fact and as aspiration. (Mills 1959, pp. 170–2)

Gorman (1977, p. 83) contrasts the Schutzian actor with another who is 'self-consciously aware' and 'critically consider[s] and evaluat[es] the circumstances (including social recipes) he or she acts in'. Such a conception is closer to that of Lowe. As discussed above, Lowe explicitly rejected mindless conformity, arguing that socialisation must be flexible enough to permit the 'type to develop beyond the type'. The socialised individual of democratic society must be recognised as 'dynamic', meaning 'incomplete' (that is, still growing, changing, developing, learning), and the individual must never uncritically accept society's prevailing standards and ideas.

Lowe recognised the potential drawbacks even of 'too much' spontaneous conformity: 'the emancipatory goal must not be conceived as a macro order from which all frictions are removed. Some degree of disorder is the price of autonomous individuation, and thus genuine emancipation' (1988, p. 13). The key is to not cross the threshold beyond which there exists a threat to the viability of society and thus the basis for individual self-actualisation. Peirce likewise noted that a considerable amount of what he termed 'chance' and 'novelty' is tolerable. For Peirce, system flexibility permits continuous 'reordering' allowing increasingly greater internal variation (Neville 1992, pp. 40–1). In fact, such flexibility is indicative of the robustness of a system.

For Lowe, the feasible maximum of individual autonomy is an ever-present goal. But it must be reconciled with 'quasi order', in other words, 'admitting some degree of disorder and instability, so long as the critical threshold, below which the persistence of society is in danger, has not been overstepped' (Lowe 1988, p. 13). The purpose of political economics, with its instrumental analysis, is to serve human society toward these ends:

> We study the structural limits of human decision in an attempt to find points of effective intervention, to know what can and must be structurally changed if the role of explicit decision in history-making is to be enlarged ... We study history to discern the alternatives within which human reason and human freedom can now make history. (Mills 1959, p. 174)

This issue to which Mills refers – the enlargement of 'the role of explicit decision in history-making' recalls Lowe's earlier remark to the effect that 'From now on, the future will have to be more and more the result of our deliberate choices, at every level of human activity,' analogous to the tendency in the development of the individual from childhood to adulthood for processes of habit-change and norm-adoption to become increasingly self-determined. I want to also relate it to another phenomenon emphasised by Lowe cited earlier regarding 'the 'incompleteness' of human history, and ... the significance of cultural evolution'.

Lowe often expressed the view that the determinism in the socioeconomic systems portrayed by the classical economists (here meaning the classical political economy of Smith and Ricardo, not neoclassical economics), exemplified by 'laws' – the 'iron law of wages', the 'law of population', 'the (classical) law of diminishing returns' – that were seen to govern relations between such factors as population (labour supply), subsistence (wages), natural resources, employment, and technical change, was rooted in the fact that the social economy of the classical era was characterised by 'impersonal forces or "laws" which might be observed or interpreted, but which could not be altered' (Lowe 1971: 568). But scientific and technological advance later transformed most of these law-like relations into variable ones, capable of human control: 'That which once "happened", can now be made to happen, or prevented from happening' (ibid.: 568). Furthermore, having created the technological potential to both induce and prevent disaster, humankind has 'no alternative to accepting the challenges of the new era':

> In the face of this tremendous enlargement of human capabilities, there is no possibility of turning away. Even doing nothing, or outlawing the advance of further capabilities, would be as much an act of intervention as exploiting our newfound capabilities to the utmost. (Lowe 1971: 368)

I want to argue that this all has relevance for the issues of spontaneity and intentionality, and of *cosmos* and *taxis*, with which we are concerned. First, there may have been a time when, relatively speaking, there was little we could do about the negative unintended effects of our socioeconomic system. This is no longer the case, or at least much less so. With our new capabilities come new responsibilities from which 'there is no possibility of turning away'. In confronting the challenges of poverty and unemployment in the 21st century, we not only can draw on the tools of new technologies and new knowledges – we can also utilise the strategies and tactics for creative discovery outlined herein to devise innovative approaches to policy formulation and implementation. Furthermore, the enhancement of the powers of creative discovery and imaginative problem-solving are possible not only for entrepreneurs, scientists, physicians, detectives, and policy-makers, but for all citizens and members of the community, and should be part of the educational curriculum. The extension and expansion of the realm of creative discovery and imaginative problem-solving is also part of a transition in the evolutionary process from one of unconscious evolution to conscious evolution of the human species, and not just conscious evolution, but the evolution of consciousness (Roszak 1975). Thus, much of what was once 'spontaneous' is becoming increasingly intentional, can become increasingly intentional, and must become increasingly intentional, while conscious efforts for individual and social betterment – rooted in individual and social responsibility – can be increasingly experienced as 'spontaneous' decision as the result of conscious emphasis on empathetic understanding and the creative process in education and other sites of socialisation. In this way, the communal basis for individuation and self-actualisation can be strengthened, and the capacity for individual and social freedom enlarged.

## NOTES

1 Of course, the voluntary adoption of social codes or rules of conduct is by no means isolated to England, a point that Lowe emphasises: 'It is true I have spoken of England. But in the last resort I am not concerned with the unique historical phenomenon of a particular country and people and their future fate, but with the general pattern of a society whose mode of life is spontaneous collectivism' (1937a, p. 40).
2 For more on Vygotsky and Lowe, see Forstater (1997: 161).
3 For relevant discussions of Lowe and Schutz, see Forstater (1997; 2001).
4 See for example Lowe (1969, pp. 183–4 and n24; 1976, p. 13n15; [1965] 1977, p. 332; 1992, pp. 326–7). See also Forstater (1999: 9–11).
5 The interpretation given here of 'spontaneous conformity' as related to imaginative self-control in the context of social interaction may be related to the creative responsibility entailed in the individual musician-ensemble relation in improvisational music, for example, in the African Continental and Diasporan traditions. The ability of musicians to play 'free' and yet 'together' has been attributed to the 'inner pulse control' of the individual performers (Thompson 1983, p. xiii). The greater the musicians' ability to keep the inner pulse, the freer they are to explore the furthest edges of the organising principle of the composition without things 'falling apart'. In the case of spontaneous conformity in social life, the greater the commitment to community, the greater the individual autonomy possible without social disruption.
6 Schutz takes social structure as pre-given and fixed ('imposed relevances' in his terminology), concentrating instead almost exclusively on the determination of human action within that context. 'Although Schutz recognises the institutionalisation of action in social settings, the objectification of human intentions in sign systems and language, as well as the objectivated results of human acts, he appears to consistently avoid analysing their objective basis, viewing the latter as not a vital part of his investigation' (Smart 1976, pp. 98–9).
7 'And now our basic definitions of society and of self are being overtaken by new realities. I do not mean merely that never before within limits of a single generation have men been so fully exposed at so fast a rate to such earthquakes of change. I do not mean merely that we feel we are in a kind of epochal transition, and that we struggle to grasp the outline of the new epoch we suppose ourselves to be entering. I mean that when we try to orient ourselves – if we do try

– we find that too many of our old expectations and images are, after all, tied down historically: that too many of our standard categories of thought and of feeling as often disorient us as help us explain what is happening around us; that too many of our generalisations are derived from the great historical transition from the Medieval to the Modern Age; and that when they are generalised for use today, they become unwieldy, irrelevant, not convincing. I also mean that our major orientations – liberalism and socialism – have virtually collapsed as adequate explanations of the world and ourselves' (Mills 1959, p. 166).

# REFERENCES

Accardo, Pasqualle (1987), *Diagnosis and Detection: The Medical Iconography of Sherlock Holmes*, Teaneck, NJ: Fairleigh Dickinson University Press.

Allen, Richard T. (1998), *Beyond Liberalism*, New Brunswick, NJ: Transaction.

Austin, James H. (1978), *Chase, Chance, and Creativity: The Lucky Art of Novelty*, New York, NY: Columbia University Press.

Berk, L. E. (1994), 'Vygotsky's Theory: The Importance of Make-Believe Play', *Young Children,* **50** (1): 30–9.

Clary, Betsy Jane (1998), 'Lowe and Tillich on the Church and Economic Order', in Harald Hagemann and Heinz D. Kurz (eds), *Political Economy in Retrospect: Essays in the Memory of Adolph Lowe*, Cheltenham: Edward Elgar, pp. 257–82.

Dewey, John (1916), *Democracy and Education*, New York, NY: Macmillan.

Forstater, M. (1997), 'Adolph Lowe and the Austrians', *Advances in Austrian Economics* **4**: 157–73.

Forstater, M. (1999), 'Working Backwards: Instrumental Analysis as a Policy Discovery Procedure', *Review of Political Economy*, **11** (1): 5–18.

Forstater, M. (2001), 'Phenomenological and Interpretive-Structural Approaches to Economics and Sociology: Schutzian Themes in Adolph Lowe's Political Economics', *Review of Austrian Economics,* **14** (2/3): 209–18.

Gelwick, Richard (1977), *The Way of Discovery*, New York, NY: Oxford University Press.

Gorman, Robert (1977), *The Dual Vision: Alfred Schutz and the Myth of Phenomenological Social Science*, London: Routledge & Kegan Paul.

Hotson, L. (1942), 'Literary Serendipity', *ELH*, **9** (2): 79–94.

Lowe, Adolph (1935), *Economics and Sociology: A Plea for Cooperation in the Social Sciences*, London: George Allen & Unwin.

Lowe, Adolph (1937a), *The Price of Liberty*, London: Hogarth.

Lowe, A. (1937b), 'The Task of Democratic Education: Pre-Hitler Germany and England', *Social Research,* **4**: 381–98.

Lowe, Adolph (1940), *The Universities in Transformation*, London: Sheldon Press.

Lowe, A. (1942), 'A Reconsideration of the Law of Supply and Demand', *Social Research*, **5**: 431–57.

Lowe, A. (1951), 'On the Mechanistic Approach in Economics', *Social Research*, **18**: 403–34.

Lowe, Adolph ([1965] 1977), *On Economic Knowledge: Toward a Science of Political Economics*, Enlarged edition, Armonk: M. E. Sharpe.

Lowe, Adolph (1969), 'Economic Means and Social Ends: A Rejoinder', in Robert L. Heilbroner (ed.), *Economic Means and Social Ends: Essays in Political Economics*, Englewood Cliffs, NJ: Prentice-Hall, pp. 1–36 and 167–99.

Lowe, A. (1971), 'Is Present-Day Higher Learning "Relevant"?', *Social Research*, **38**: 563–80.

Lowe, Adolph (1976), *The Path of Economic Growth*, Cambridge: Cambridge University Press.

Lowe, Adolph (1988), *Has Freedom a Future?*, New York, NY: Praeger.

Lowe, Adolph (1992), 'A Self-Portrait', in Philip Arestis and Malcolm Sawyer (eds), *A Biographical Dictionary of Dissenting Economists*, Aldershot: Edward Elgar, pp. 323–8

Mills, C. Wright (1959), *The Sociological Imagination*, New York, NY: Oxford University Press.

Neville, Robert Cummings (1992), *The Highroad Around Modernism*, Albany, NY: SUNY Press.

Ochs, Peter (1993), 'Charles Sanders Peirce', in David R. Griffin, John B. Cobb, Jr., Marcus P. Ford, Pete A. Y. Gunter and Peter Ochs (eds.), *Founders of Constructive Postmodern Philosophy: Pierce, James, Bergson, Whitehead, and Hartshorne*, Albany, NY: SUNY Press, pp. 43–88.

Peirce, Charles S. (1931–1935), *Division of Signs, Collected Papers of Charles Sanders Pierce, Vols. I–IV*, Charles Hartshorne and Paul Weiss (eds), Cambridge, MA: Harvard University Press.

Polanyi, Michael (1958), *Personal Knowledge*, Chicago, IL: University of Chicago Press.

Polanyi, Michael (1959), *The Study of Man*, Chicago, IL: University of Chicago Press.

Polanyi, Michael (1966), *The Tacit Dimension*, Garden City, NY: Doubleday.

Polanyi, Michael ([1966] 1997), 'Creative Imagination', in Richard T. Allen (ed.), *Society, Economics & Philosophy: Selected Papers of Michael Polanyi*, New Brunswick, NJ and London: Transaction, pp. 249–66.

Polanyi, Michael ([1972] 1997), 'Genius in Science', in Richard T. Allen (ed.), *Society, Economics & Philosophy: Selected Papers of Michael Polanyi*, New Brunswick, NJ and London: Transaction, pp. 267–81.

Polya, George ([1945] 1957), *How To Solve It: A New Aspect of Mathematical Method*, Second edition, Princeton, NJ: Princeton University Press.

Polya, George ([1948] 1984), 'On Patterns of Plausible Inference', in Gian-Carlo Rota (ed.), *George Polya: Collected Papers, Volume IV: Probability, Combinatorics; Teaching and Learning in Mathematics*, Cambridge, MA: MIT Press, pp. 473–484.

Polya, George ([1949] 1984), 'Preliminary Remarks on the Logic of Plausible Inference', in Gian-Carlo Rota (ed.), *George Polya: Collected Papers, Volume IV: Probability, Combinatorics; Teaching and Learning in Mathematics*, Cambridge, MA: MIT Press, pp. 488–495

Polya, George ([1952] 1981), *Mathematical Discovery: On Understanding, Learning, and Teaching Problem Solving*, New York, NY: Wiley & Sons.

Polya, George ([1971] 1984), 'Methodology or Heuristics, Strategy or Tactics?', in Gian-Carlo Rota (ed.), *George Polya: Collected Papers, Volume IV: Probability, Combinatorics; Teaching and Learning in Mathematics*, Cambridge, MA: MIT Press, pp. 586–594.

Roszak, Theodore (1975), *Unfinished Animal*, New York, NY: Harper & Row.
Schutz, Alfred ([1943] 1970), 'The Problem of Rationality in the Social World', in Dorethy Emmet and Alasdair MacIntyre (eds), *Sociological Theory and Philosophical Analysis*, New York, NY: Macmillan, pp. 89–114.
Schutz, Alfred [1953] (1967), 'Common-Sense and Scientific Interpretation of Human Action', in Maurice Natanson (ed.), *Collected Papers, Volume I – The Problem of Social Reality*, The Hague: Martinus Nijhoff, pp. 3–47.
Smart, Barry (1976), *Sociology, Phenomenology, and Marxian Analysis*, London: Routledge & Kegan Paul.
Thompson, Robert Farris (1983), *Flash of the Spirit: African and Afro-American Art and Philosophy*, New York, NY: Random House.
Vygotsky, Lev S. ([1934] 1962), *Thought and Language*, Cambridge, MA: MIT Press.

# 10. The laboratory and the market – on the production and interpretation of knowledge

**Hans Siggaard Jensen and Lykke Margot Richter**

Scientific knowledge is becoming increasingly important economically. Knowledge about science is also. There is a long tradition of understanding science under the auspices of philosophy, and a shorter tradition of viewing science as a social activity. Science has been made into a model for the rational, the quintessential creation of modernity. But is science a creation like a building, a construction, or is it, like language, the market or artistic styles, a result of action but not of design? What kinds of order are produced in science and what are the relationships between ways of producing knowledge and different types of orders – if any? These are questions we will discuss in the present chapter.

## 1. SCIENTIFIC DISCIPLINE

For a very long time the ideal order of a scientific theory has been the order of a deductive system. Euclid's *Elements* have been seen as the first good example (Heath 1908). A database can give us information, and we can know what type of information is contained in it, but is it knowledge? In the history of philosophical thought the question regarding knowledge has been seen as the problem of defining the necessary and sufficient conditions for a proposition; furthermore, also to draw a distinction between what is believed to be true and what is actually proven to be true belief, the latter in order to call it knowledge. And so scientific knowledge is about theories and explanations, which involves at least deductions, or is it?

Rationalists have long seen the Euclidian-Cartesian model of knowledge as the ideal. We only need assumptions – ideally self-evident assumptions – and logical deductions from these and we are in the business of producing knowledge. The knowledge produced then takes the form of a deductive system, with all the complexity of such a system as evidenced by the study of such systems in mathematical logic. Yet as soon as they allow for the statement of identities (using the '=' symbol) they are so complex that the concepts of deductive proof and model-theoretic truth do not coincide, as Gödel ([1931] 1962) discovered with his incompleteness theorem:

> (1): If T is a sufficiently powerful formal theory and T is sufficiently sound, then T is incomplete, i.e. there are true sentences undecided by T.

and

> (2): If T is a sufficiently powerful formal theory, then T cannot prove its own consistency.

Through a complex social process deductive systems are made to unfold – the work of mathematicians. Various branches of mathematics are developed and attempts are made at unification. The creation of set theory at the turn of the 20th century is an example; or the creation of category theory in the middle of the century. Two different ways of getting an 'overview' of mathematics – either a common language – sets, relations, functions – or a common theory in which the various theories can be 'located' – this or that part of mathematics is really a such-and-such type of category. We could in a simple model say that a deductive database and its deductive closure are what this is all about. We seldom retract statements when they have been proved. So in a way we expand all the time. Falsifications are not part of mathematics. This is one way in which it is not empirical. As soon as we have an empirical element then the situation gets more complicated. Then we need the work of the laboratory, the work of the observatory, the scientific expedition or the clinical test. We need to assess the relation between empirical evidence and theories. This is done in many different ways, exemplifying the so-called methods of scientific investigation. The work is done both in the laboratory – taking it to be paradigmatic – and through communication of results to a group of people. The result is what might be called a discipline – or a science. It consists of textbooks, journals and so forth, all accumulated knowledge. Of course we often here find retractions, that is we discover that what we believed was knowledge was actually only belief. Ideally again we try to model a discipline as a deductive system. We can say that it is a dynamical deductive system in which there is a wish to obtain consistency, but this might be an ideal that is never attained. So we add propositions and obtain inconsistencies and thus need to retract others. The literature on belief revision exemplifies this type of modelling of what has been called 'the flux of knowledge' (Gärdenfors 1988). The function of empirical evidence is thus to be part of revisions of our assumptions and at the same time add power to our deductions. Thus evidence can decide the issue between two or more theories, or it can add to our deductive system. Thus the sciences – our knowledge – can be modelled as a number of deductive systems. As is well known it was an aim of logical positivism to try to unify these systems into one. As is also known this was never achieved. To give meaning to our deductive systems we need more than just formalisms, we need analogies and metaphors, such as we find in the non-formal models in scientific theories. How is knowledge useful if understood in this way? Basically the way it is useful for non-theoretical purposes, that is purposes that have nothing to do with explanations in the theoretical sense, is – in the paradigmatic case – through statements formulating causal laws.

> To give a causal explanation of an event means to deduce a statement which describes it, using as premises of the deduction one or more universal laws, together with certain singular statements, the initial conditions. (Popper [1934] 1959, p. 87–8)

They have the logical form of if-then statements, and thus allow for deductions. If one wants to obtain a certain state of affairs of type A and there is a law saying that 'If a state of affairs of type B obtains then at a later time a state of affairs of type A will come to obtain' (B causes A), then we only need to produce a type-B situation to get A. Thus we need the knowledge of the causal law and the necessary knowledge and ability to produce an A-type situation. This of course involves further knowledge and abilities, powers of action and foresight, and of course also the ability to classify and identify types of situations or states of affairs. We are not saying that the only form of knowledge that is applicable is the one expressing causal laws; functionally expressed knowledge can of course also be applied, but then through the control of parameters, making it a type of instantaneous 'casual' law. If $k = pv$ then by controlling volume pressure can be controlled or vice versa.

The result of knowledge production, which is a social process of cooperation and communication, is then a set of systems of statements organised deductively. To interpret the system of course one needs models, analogies and metaphors. Some of these are explicit, others implicit. We can make an analogy to a set of buildings (maybe even a city), where the result of cooperation and communication is a static set of structures. The best available concrete examples are libraries and databases. A good philosophical example would be Popper's so-called World Three model of knowledge, the result of productions shaped by evolutionary forces (Popper 1972). What is in World Three is the result of a long process of trial and error, of the unfolding of human reason and rationality. This model of scientific research then has the following features. The sciences are organised as disciplines, and disciplines are progressively obtaining the shape of deductive systems of greater and greater scope. Disciplines are defined by the fundamental assumptions and of course by partitioning of phenomena, which again is the result of certain fundamental assumptions. Problems of a theoretical nature arise inside disciplines, and are basically solved in the way that mathematical problems are solved, although observations can acquire the status of statements, and thus be part of the deductive system. There is a distinction between basic science and applied science, where applications are linked to solutions of non-theoretical problems. The disciplinary world is a world of isolation from practice, ideally a closed world modelled formally. The relation to the actual world is secured through a general hypothesis of uniformity, and the processes of abstraction and idealisation. Measurements establish a relation between observed properties and numbers (of some sort), and are thus understood not as proportions but as functions. Applied disciplines are basically the application of knowledge through its transformation into causally effective forms, that in the forms of rules can guide action. Applied disciplines are thus a theoretical mix of causal principles and heuristics that can establish best practices, for instance through forms of optimisation. The applied disciplines do not have any clear deductive order. The actual practice based on theory is a mixture of knowledge and skill, where skill can be to a large extent craftsman-like in character. The relation between a discipline and a profession is then a homogenous relation, where the professional is scientifically trained and therefore knowledgeable and has 'learned' to use the knowledge in and through practice and has thus become skilled. The persons attached to the discipline (the botanists, astronomers and so forth) can be organised in different ways but common to all are the principles of self-governance – a closing principle – and self-administration of a specific ethos concerning quality and conduct. This is usually called the 'peer'-system. Professionals attached to the discipline are often called experts. We can use an analogy like the following: a group of people is in cooperation weaving a huge and complicated carpet, all working together on the same project, all totally absorbed by the project, and the carpet is the end result. Nobody knows the size of the final carpet, so it is an unending project. There may be other carpet-projects around of similar sort, but with other patterns or pictures. It is not necessarily clear to the persons producing the carpet what they are doing in the sense that they can describe and analyse it, but they can produce. It may also sometimes happen that they have to start anew with certain parts, so that small or large parts of the hitherto-produced carpet have to be discarded. In the case of science these two situations are analogous to the points made by Latour and Woolgar and by Kuhn about science produced within the framework of traditions and habitual routine practices (Latour and Woolgar 1981; Kuhn 1970).

## 2. TYPES OF KNOWLEDGE

The type of knowledge thus far described is typically produced in what has been termed the Modus 1 type of knowledge-production system. This is characterised by the following:

- Problems arise and are solved within well-established disciplines.
- Such disciplines have clear paradigms to work within.
- A linear relation is aimed at between theoretical development and practical problem-solving, between basic and applied research.
- There is a homogenous relation between theoretical experience, qualification and practical experience, which means that there is the possibility of a one-to-one relation between a discipline and a profession.
- The concept of quality is internal and oriented towards the discipline and functions via 'peers'.

In the book *The New Production of Knowledge* – from which the Modus 1 characteristics are drawn – it is claimed that a new type of knowledge-production system is emerging, Modus 2 (Gibbons, Scott, Nowotny, Limoges, Schwartzmann and Trow 1994; and see also Nowotny, Scott and Gibbons 2001). The system is characterised by the following:

- Problems arose and are formulated in the context of application.
- The production of knowledge does not take place primarily within the framework of a discipline, but is trans-disciplinary.
- Communication and application of knowledge take place in the context of production.
- The relation between theory, experience and qualification is heterogeneous.
- The criteria of quality are more dependent on social relevance and utility than on an intra-disciplinary context.

The areas of scientific activity where we find Modus 2 knowledge-production are areas like software, systems management, and biotechnology. In a way it can be said that the sciences of the artificial or of design are paradigmatic whether it be the study of technical or social artefacts. Thus whereas Modus 1 is typically found in the natural sciences, Modus 2 is found in the technical and social sciences (maybe also the human sciences if we consider language and works of art as types of artefacts). There may be parts of the social and human sciences were Modus 1 traits are dominant, such as parts of economics and linguistics.

Let us look at the type of activity that would be epistemically relevant in understanding the processes and products of Modus 2 knowledge production. We may as a beginning say that it is very close to creating knowledge by learning from experience, but not the experience of working in the laboratory. The knowledge created is created in a different type of interaction with the object of knowledge, which is sometimes even created in the same process. So for instance we get knowledge about software in the process of producing it. In a way we could call this type of knowledge poetic in the original sense. The production of knowledge out of experience in this sense involves conceptualisation through classification and creations of typologies, analysis and interpretation, and the formulation of explicit rules and heuristics. Explanations may mimic causal explanations, as when we explain actions through norms or rules or types of institutions; and logical deductions are certainly not excluded. But we may say that the process is a process of clarification, where the object of knowledge is constructed as part of the process. Often the process is described as a process of transformation, transformation from implicit or tacit to explicit. But of course the transformation is also a process of translation and thereby of creation. The problem with the idea of a transformation of course is that knowledge is already in some sense present, but in a different form, and also in many forms. For when we study the work of the baker (he who in Nonaka's terms has tacit knowledge (Nonaka 1991)), and writes it down in words, the tacit knowledge is not transformed, it is still there – now we just have a new type of knowledge that relates to the work of the baker. In this conception of knowledge a fundamental aspect of knowledge is its dynamic character. It is only knowledge when it is used or is in circulation. In this sense knowledge is a practice and not the result of a practice. This in some way also then questions the distinction between the content and the methods of science. The social character of Modus 2 knowledge-production gives it a very different character from that found in Modus 1. The idea of knowledge as a deductive system is totally improper. Knowledge is much more a set of narratives that gives guidance concerning practices: narratives that in this case contain forms of knowledge – models, analogies, and metaphors. This points to the fact that Modus 1 types of knowledge can only get meaning – the formalisms can only be interpreted – through models, analogies and metaphors. And it can only be applied through skills that have the character of heuristics or rules that are understood and are part of social practices. Some of these practices can be made explicit, formulated, conceptualised. But then we engage in Modus 2 knowledge-creation. Modus 2 knowledge-production is a process, and the result is not necessarily structured like a deductive system. Does it have a structure at all? This is like asking how the meanings are created when meanings are created and structured. We could conceivably try to respond with at least three different types of structure. There are linear structures where the part-whole relationship is such that there exist parts that are more simple than the whole; there are recursive structures where parts can produce wholes, and thus part and whole have the same complexity; and there are fractal structures where parts are more complex than wholes. Deductive structures are recursive because it is possible to see deduction as calculation and ultimately as computation. Most scientific theories when formulated as mathematical are linear. We look for simple parts. Classical mechanics and neoclassical economics are good – and related – examples; look at the Appendix to Jevons ([1871] 1970) *The Theory of Political Economy* as an example. Practices and meanings are recursive or fractal. If a structure is contextual we have a fractal structure because the part can only get a meaning as part of a whole, so really understanding the part involves more than the part – this is a sort of holism. We can delineate but not describe the part, yet not understand it as what it is in itself. Recursive accounts of meaning and practices see them as a sort of grammars, generative structures. We can say that paradigmatically within Modus 1 knowledge, what can be termed knowledge of Type 1 is linear, whereas knowledge of Type 2 is fractal.

## 3. ON PRODUCTION AND INTERPRETATION OF KNOWLEDGE

When we look at the relation between production, creation/construction and discovery in relation to knowledge we can also find interesting differences between the two types of knowledge. Type 1 knowledge is typically understood as discovered – both as the result of observation and experiment and of deduction. Knowledge is either a part of the natural world being 'out there' to be discovered, or existing in a platonic world (the interesting question about the relation between mathematics and empirical knowledge expressed mathematically/formalised is a result). Knowledge is knowledge about the facts and the world contains facts, and the production of knowledge is the discovery of facts formulated in a language (typically the language of mathematics). Often we suppose facts to be identical with (or isomorphic with) propositions. Statements that are true express – whatever that means – or refer to propositions and thereby to facts. An important question then of course is how to account for theoretically basic statements (are they true like statements that refer to or express propositions or in some other way?). But we can say that knowledge is preexisting, as facts (empirical or deductive facts – deductive facts are logical connections); and the production of knowledge is a discovery process where facts are formulated or expressed. But the facts of course exist – on this conception – independently of their formulation or expression. So the world is considered a place full of facts that are lying around 'ready' to be discovered. The productive activity is then one of experience in the sense of observing and discovering, and a process of formulation that allows explanation and eventually control. This is mainly done through the equivalent logical form of a simple deduction and a causal connection, like this:

> If a and $a \rightarrow b$ then b;
>
> If A is the case and always: if A is the case then B will be the case, then B will be the case

This allows a parallelism between proof of b and explanation of B, in the causal case but also in the functional, like:

> If p has value a and $k = pv$ (and $k=c$) then $v=b=c/a$

Where the functional connection is that pressure times volume is a constant (of course a simplified example but in principle like all other functional connections). The basic problem of connection between mathematical structures and empirical facts is the determination of the values of variables, measurement, and the classification into types of situations. Because in the causal case there is really talk about a particular situation and that situation being of a specific type, so we should actually say something like this:

> If A is the case, and always: if a situation of the same type as A is the case then a situation of type B will be the case, then a concrete instance of B will be the case

If we look at this we clearly see that the conditions for the description of types of situations and of concrete situations are extremely important, just as measurements are important. And both measurement and descriptions of types involve basic conceptualisations, classifications and maybe even some basic theories.

In the case of knowledge of Type 2 we do not conceive of facts, propositions and knowledge in the same way. Let us for a moment accept the term 'tacit knowledge', because then it is clear that when we do, concepts like 'tacit belief' or 'tacit information' will give problems. Tacit knowledge is somehow already secure as knowledge. So it already exists, but it is – because it is tacit – not in any way structured like propositions or facts; it is more like a disposition to perform certain acts – including speech acts – in certain situations. Making it explicit is then like creating a set of typologies and rules both describing and prescribing the dispositions and the actions, like this:

If in situation $S_1$ a new situation $S_2$ is desired do actions $A_1$ .... $A_n$

We now know how to get a situation of the same type as $S_2$. So, if we have this desire and the knowledge and we are in a situation like $S_1$ we have a disposition to act like $A_1$ ... $A_n$. We can also interpret $S_2$ as the results of desires and knowledge and similar typologies like the one we have. We can say that this type of knowledge is not only poetic, but in a way also rhetorical in the sense that it is concerned with bringing about certain states of affairs. Of course this can involve causal knowledge, but basically we would rather use knowledge about certain powers based on experience. This is something like knowing the efficacy of rules; I know that if I do or say this then somebody else will do so and so and therefore a certain state of affairs will come to obtain (with reasonable probability or likelihood). This both involves being able to follow rules, and the social conditions for such rule-following, and considerations of efficacy, which means that we are concerned about whether we get the intended results or not. Tacit knowledge as a concept is often seen as originating in the work of Michael Polanyi (Polanyi 1958). He most often used the term 'tacit knowing', making a sharp distinction between knowledge and knowing. Knowing is the subjective and personal aspect that has process-character, whereas the term knowledge signifies that something more substantial is present.

Basically, we can say that Type 1 knowledge is the type of knowledge that concerned Descartes and Locke (Descartes [1637] 1983, Locke [1671] 1975). It is the type of knowledge that became of central importance with the emerging institution of natural and experimental science, in contradistinction to revealed knowledge or knowledge as a result of logical argumentation (which again is different from mathematical knowledge). Knowledge of Type 2 can be considered the type of knowledge that concerned Dewey and

Wittgenstein, knowledge as the result of inquiry-processes and knowledge as embedded in social institutions and forms of life (Dewey 1938, Wittgenstein 1953).

Producing knowledge of Type 1 is striving through the intellectual work of persons in a community – the scientific community – to obtain a non-personal detached type of knowledge. It is abstract in this sense and it has the character of an objective product. Not only is objectivity a desired epistemological trait of this kind of knowledge, it also has the sense of trying to become like an object, so that knowing it is just like perceiving an object – like the metaphor of 'seeing the facts'.

Producing Knowledge of Type 2 is different, although no less a social process. But the knowledge produced and the production process are difficult to separate. The difference is like describing the market at a general equilibrium and trying to prove the existence of such an equilibrium – typically by looking at a representation of the market as a set of equations, and describing the market as Hayek does it in terms of discovery procedures (Hayek 1978). Through the work of the laboratory and its conception of discovery of facts through communication and exchange, an objective structure is produced – knowledge. The type of work connected with producing knowledge of Type 2 is much more like the embedded and situated position that we find ourselves in when we are agents acting in a market. Maybe one could say that the knowledge of Type 1 is like the knowledge an economist tries to get about the market through the investigation of a model of the market, and the knowledge of Type 2 is like the knowledge an agent on the market gets. To formulate a relevant model one would need at least to be able to formulate parts of the knowledge one gets through operating in the market.

## 4. CLOSURE

We have outlined and discussed some problems of the forms of knowledge related to types of structures. We find such structures both in the intellectual and the social sphere. In the present situation where the role of knowledge in society is changing and the forms of the production of knowledge also, it seems important to understand theoretically the interplay between social and intellectual structures.

## REFERENCES

Descartes, René ([1637] 1993), *Discourse on Method and Meditations on First Philosophy*, tr. by Donald A. Cress, Third edition, Indianapolis, IN: Hackett.

Dewey, John (1938), *Logic: The Theory of Inquiry*, New York, NY: Henry Holt and Company.

Euclid's Elements, *The Thirteen Books of Euclid's Elements*, (1908) tr. by T. L. Heath, Cambridge: Cambridge University Press.

Gibbons, Michael, Peter Scott, Helga Nowotny, Camille Limoges, Simon Schwartzmann and Martin Trow (1994), *The New Production of Knowledge: The Dynamics of Science and Research in Contemporary Societies*, London: Sage.

Gärdenfors, Peter (1988), *Knowledge In Flux: Modeling the Dynamics of Epistemic States*, Cambridge, MA: MIT Press.

Gödel, Kurt ([1931] 1962), *On Formally Undecidable Propositions of Principia Mathematica and Related Systems*, Edinburgh: Oliver & Boyd.

Hayek, Friedrich A. von (1978), 'Competition as a Discovery Procedure', in Friedrich A. von Hayek (ed.), *New Studies in Philosophy, Politics, Economics and History of Ideas*, London: Routledge & Kegan Paul, pp. 179–90.

Jevons, W. Stanley ([1871] 1970), *The Theory of Political Economy*, Harmondsworth: Penguin Books.

Kuhn, Thomas S. (1970), *The Structure of Scientific Revolutions*, Second edition, Chicago, IL: University of Chicago Press.

Latour, Bruno and Steve Woolgar (1981), *Laboratory Life – The Social Construction of Scientific Facts*, Second edition, Beverly Hills, CA: Sage.

Locke, John ([1671] 1975), *An Essay Concerning Human Understanding*, Oxford: Clarendon Press.

Nonaka, I. (1991), 'The Knowledge-Creating Company', *Harvard Business Review*, **69** (6): 96–104.

Nowotny, Helga, Peter Scott and Michael Gibbons (2001), *Rethinking Science: Knowledge and the Public in an Age of Uncertainty*, Cambridge: Polity Press.

Polanyi, Michael (1958), *Personal Knowledge – Towards a Post-Critical Philosophy*, Chicago, IL: Chicago University Press.

Popper, Karl R. ([1934] 1959), *The Logic of Scientific Discovery*, London: Hutchinson.

Popper, Karl R. (1972), *Objective Knowledge – An Evolutionary Approach*, Oxford: Clarendon Press.

Wittgenstein, Ludwig (1953), *Philosophical Investigations*, Oxford: Basil Blackwell.

# Index